Demanding Energy

Allison Hui • Rosie Day • Gordon Walker
Editors

Demanding Energy

Space, Time and Change

palgrave
macmillan

Editors
Allison Hui
Department of Sociology
Lancaster University
Lancaster, UK

Rosie Day
School of Geography Earth and
Environmental Sciences
University of Birmingham
Birmingham, UK

Gordon Walker
Lancaster Environment Centre
Lancaster University
Lancaster, UK

ISBN 978-3-319-61990-3   ISBN 978-3-319-61991-0   (eBook)
DOI 10.1007/978-3-319-61991-0

Library of Congress Control Number: 2017954396

Cover illustration: xijian/Getty

Printed on acid-free paper

This Palgrave Macmillan imprint is published by Springer Nature
The registered company is Springer International Publishing AG
The registered company address is: Gewerbestrasse 11, 6330 Cham, Switzerland

Preface

One of the core interests of this book is to situate discussions of energy demand and the performances and practices that constitute it within particular spaces and times. It would be remiss of us then if we didn't similarly situate the book itself.

Demanding Energy grew from and was supported by the nexus of practices around the DEMAND Centre (Dynamics of Energy, Mobility and Demand), a multi-institutional research collaboration funded by the Engineering and Physical Sciences Research Council (grant number EP/K011723/1) as part of the Research Councils UK (RCUK) Energy Programme and by Électricité de France (EDF) as part of the R&D European Center and Laboratories for Energy Efficiency Research (ECLEER) Programme. Established in 2013 and co-directed by Elizabeth Shove and Gordon Walker, the centre set out with three core propositions: (1) that energy demand is an outcome of social practices, (2) that such practices are shaped by institutions and infrastructures and (3) that these arrangements reproduce historically and culturally specific interpretations of need and entitlement. These propositions became the basis for a wide range of empirical projects around four main themes: trends and patterns in energy demand, how end use practices change, managing infrastructures of supply and demand, and normality, need and entitlement.

Inspired by a range of creative and engaging intellectual and social exchanges featuring reading groups, seminar series, writing exchanges around 'pieces of thought', annual 'clan gatherings', floorball, photo competitions, birthday cake, cooking and recipe sharing (see www. demand.ac.uk), authors in this collection were able to develop a novel, shared approach to understanding what energy is for. Those authors who were not a part of the DEMAND Centre itself joined in with the discussions as participants in our programme of international visitors or co-presenters at conferences, including the DEMAND International Conference in April 2016 (http://www.demand.ac.uk/conference-2016/). We are indebted to all of those colleagues, and particularly other DEMANDers, who shared in these practices with us—shaping our trajectories of analytic and theoretical development as well as ensuring our exchanges were sometimes unconventional and as often as possible fun.

In between the discussions that sparked these chapters and the discussions that we hope emanate from them lies the text itself. Many thanks to Oliver Fitton for his work and innumerable tracked changes, emails, draft documents and pieces of feedback during the compilation of the manuscript.

Lancaster, UK   Allison Hui

                                                                   Gordon Walker
Birmingham, UK   Rosie Day

Contents

List of Figures

List of Tables

1

Demanding Energy: An Introduction

Allison Hui, Rosie Day, and Gordon Walker

The café is quiet this morning. Just a few customers taking their time: a couple with suitcases stopping off on their way to the railway station and a guy settling in with his pastry and laptop. I like working here when it's this empty—my thoughts wander in interesting ways. It's a good place for trying out new ideas. Whenever I can, I get here early to make sure I can get a table and plug in whatever needs charging. And sit away from the heaters and the speakers—some of the music's ok, but not all of it. Most mornings, I reckon there are more staff here than customers. New ones keep appearing—fetching boxes from the store room, looking at clipboards, bringing toppings to make up sandwiches for later in the day. Can't work out how their shifts are arranged. There are a few more customers coming in now, coffees to be made, money exchanged. I wonder if anyone will have time to change that burned out light bulb.

A. Hui (✉) • G. Walker
Lancaster University, Lancaster, UK

R. Day
University of Birmingham, Birmingham, UK

© The Author(s) 2018
A. Hui et al. (eds.), *Demanding Energy*, DOI 10.1007/978-3-319-61991-0_1

Accounts of energy demand can start from many places. Physicists might begin with the first law of thermodynamics, the rule that energy cannot be created or destroyed. Economists might start from models that forecast future trends in relation to economic growth. Utility company workers might privilege the meters whose readings slowly measure accumulating flows of gas or electricity. Bus company managers might emphasise the route changes that become necessary as diesel vehicles are retired to make way for electric ones. Policymakers might foreground the emissions targets and international agreements on climate change that (perhaps) make managing energy demand a priority.

In this book, we start from somewhere else, from an interest in what energy is for (Shove and Walker 2014; Walker 2014). Our core question is a provocatively expansive one and up to this point remarkably understudied: what social processes constitute and make energy demand? Just as the starting points above establish different concerns and lead to different lines of investigation, so too our starting point is decisive in setting up the questions and lines of analysis that follow. By asking about the social processes that constitute energy demand, we mark out a concern for how energy demand is embedded in the shared practices and activities that make up the ongoing flow of society—such as working, commuting, eating, going to music festivals, staying in hotels, cleaning houses, visiting hospitals, walking dogs, travelling in retirement and running cafes, among many other diverse and varied things. In order to understand energy demand—its making, patterning, variation and dynamics—we argue that it is necessary to understand the practices and processes that underpin and ultimately give rise to the consumption of energy from different sources (electricity, gas, solid fuels) and the use of various energy services (transportation, heat, Wi-Fi).

We do not therefore focus on energy demand as a 'thing' with particular measurable effects, for example on supply infrastructures, or from which projections of future trends or conclusions about strategies for change might be directly drawn. Indeed, in many of the discussions in this book energy itself fades into the background as we step back from the metrics (Kw/h, ktoe: kilotonnes of oil equivalent) and energy efficiency ratings that are common foci in other accounts. Instead we foreground the diverse and varied *processes of demanding energy* that are woven into daily life (e.g. Rinkinen 2015) and contemplate how these processes have been changing over time and will continue to do so. This collection represents the first sustained effort to pose questions in these terms, develop analytic strategies and provide empirical insights that start from a concern for understanding what energy is for.

Stepping back from energy demand per se, in order to ultimately better understand it, involves focusing upon complex social relationships. There is no simple answer, for example, to the question of what or who it is that demands energy. At times it might make sense to privilege groups of people when providing explanations, or at other times technologies, and both have been studied extensively in other energy research (e.g. Isaac and van Vuuren 2009; Sahakian 2011; O'Doherty et al. 2008; Burholt and Windle 2006). Yet our starting point also makes it possible to consider how particular working practices, lifestyles, infrastructures or stages in life might be the units most consequential for processes of demanding energy.

The authors in this collection devote close attention to this range of units of analysis in order to better understand how demanding energy is a part of the practices of everyday life. For some, this involves close empirical investigations of what people are doing and how this is socially understood, an orientation informed by well-established precedents within qualitative, ethnographic and hermeneutic research. Instead of just taking for granted categories that summarise social activity—such as 'cooking', 'working' or 'home computing'—authors delve into the diversity and variation within and between such categories, engaging in detail with temporally and spatially situated enactments. Looking carefully at what people are doing and how energy becomes embroiled in these activities facilitates the challenging of assumptions about the relationship between social dynamics and energy, relationships that have been 'black boxed' or totally overlooked in previous research and policy. In addition, investigating a range of actors, materials and practices situated at particular moments, or evolving as part of historically specific transformations, allows discussion of what lies behind and before many of the summary tables, metrics, trends and load curves that are common touchstones within discussions of energy. As a result, stepping back from energy per se is shown to provide a fuller understanding of what contributes to demanding energy, and where, therefore, opportunities for change may arise. Whilst familiar energy technologies and infrastructures are ever present in these accounts, stepping back from energy per se means that they are not separated or isolated from the social worlds that they are resolutely embedded within.

In addition to this general interest in data that evidences what energy is for, and not only how much of it is used or via which technologies, some authors engage directly with theories of social practice. This body of literature provides varied accounts of how the social world is constituted by people's on-going practices, which create, sustain, transform and are influenced by diverse social structures (Giddens 1979, 1984; Schatzki 1996, 2002; Shove et al. 2012). In summarising this literature, Reckwitz emphasises how it provides a different means of approaching the role of the body, mind and knowledge in social processes (2002). Whereas economic analyses focus on single actions and choices, and other cultural theories focus on either the knowledge and meanings of the mind, the signs and symbols of texts, or the dynamics of intersubjective speech acts, practice theories focus instead upon 'blocks' of activity: "pattern[s] which can be filled out by a multitude of single and often unique actions" (Reckwitz 2002: 250). These patterned blocks are practices, a unit of study and analysis that is socially constituted, and which is shaped not only by people's actions and statements but also by socially appropriate materials, understandings, goals and procedures. It is therefore not individual dynamics that are of primary concern, but how socially shared and patterned practices are reproduced and changed. Indeed the social world can be understood as a nexus of practices (Hui et al. 2017) with particular material relations (Shove 2017; Morley 2017), power relations (Watson 2017), interconnections (Blue and Spurling 2017) and variations (Hui 2017) of relevance for thinking about energy demand.

For some authors in this volume, this theoretical orientation informed their research design, leading them to ask research questions that seek to uncover how an attention to practices, rather than, for example, the individual attitudes, behaviours and choices dominating many existing discussions of energy and sustainability (Shove 2010), can explain previously ignored or misunderstood dynamics of energy and transport demand. Others go much further, engaging in detail with specific concepts from practice theories in order to develop new empirical analyses and theoretical resources. In these cases, authors show that developing better understandings of energy demand is not about creating new categories or representations of energy demand itself, but rather about finding ways of describing and summarising social dynamics that then have important implications for how demanding energy is constituted, patterned and changing. Therefore, at the same time as they provide case-specific insights, the chapters also serve as exemplars that demonstrate and develop the scope for more sophisticated and theoretically engaged understandings of what energy is for.

By privileging processes of demanding energy, and the practices involved, this collection also challenges established boundaries within discussions of energy demand. Across the existing literature, transport-derived energy demand has largely been discussed independently from building-related (domestic and non-domestic) use and demand—as even a cursory examination of journal and book titles reveals (e.g. Inderwildi and King 2012; Williams 2012). Yet this boundary appears increasingly to be an artefact of sectoral and disciplinary boundaries, rather than an empirically sensible analytic strategy. Transformations such as flexible or teleworking, online shopping and an increase in social media apps and platforms are undoubtedly affecting where and when demanding energy occurs. Yet little evidence exists on these effects: as a 2015 UK Department for Transport report bluntly admits: "There is little or no evidence on the impact" of such developments upon travel demand (Department for Transport 2015: 66). Starting with an interest in processes of demanding energy, however, provides another way forward, as it makes any distinction between travel-derived and domestic energy demand something to question through empirical investigation rather than a pre-existing assumption. Travelling is not always easily separable from the activities that it facilitates—an insight already well established within discussions of 'derived demand' (Mokhtarian and Salomon 2001)—and therefore researching demanding energy is open to the interweaving of both. This collection therefore seeks to create greater dialogue between transport and energy literatures, by juxtaposing contributions that focus on each, as well as featuring contributions that address both together.

As well as focusing on demanding energy as ongoing processes caught up in social practices and dynamics, the subtitle of this book also indicates a concern with space, time and change. The spatial, the temporal and ongoing change are fundamental to all social processes, shaping of all relationships, ever present and always being (re)produced. It would be hard to find three more foundational concepts of social scientific enquiry. Yet the richness and depth with which they have been addressed in available socio-theoretical analyses has yet to be integrated into socio-scientific understandings of energy demand. Our decision to foreground them was therefore not only because they are foundational to social life, but also because of a need to move beyond largely implicit or limited discussions of space, time and change within existing energy demand literature. Engaging more intensively with social scientific understandings of space, time and change, we argue, provides more vivid and nuanced understandings of social practices, and ultimately of what energy is for. In the three sections that follow we therefore lay out some of this territory and particular understandings of space, time and change that can contribute to more interesting and developed analyses.

1.1   Space and Demanding Energy

All energy use evidently takes place in space and in principle can be demarcated in those terms. A television set consuming electricity can be located in cartographic terms at an address, at a point in physical space. Such individual instances of energy use can be combined into bounded spatial units—such as energy consumption within a household or within an office—units which are then amenable to aggregation into bigger ones: districts, town, regions, nations and so on. In one sense then, there is an apparently straightforward spatiality to energy demand that aligns with an understanding of space as physical, fixed and laid out across an objective surface on which the social world plays out (Massey 2005). In some cases, chapters in this volume work with this straightforward understanding of space; Durand-Daubin and Anderson, for example, contrasting cooking and eating practices in two national contexts, UK and France, although in terms of patterns of activity rather than energy use per se.

Assigning energy use to fixed locations and bounded units is not, however, without its complexities. For example, though airplanes in flight have fuelled at particular locations on the ground, their use of energy is in motion, in airspace marked by natural airstream flows and quite different air traffic control boundaries (Lin 2016). Mobile phones are similarly measured in energy consumption terms at locations of charging, yet carry their store of electricity with them so that this energy can be used for work in multiple locations, distant in space (and time) from where they were charged (Lord et al. 2015). Along with other information technologies, phones are also reliant on infrastructures of data flow and storage that can stretch across the globe, all powered and reliant on energy for their functioning (Wiig 2013). How exactly then can we locate the energy use entailed by the movement of a text message—in the sending device, the receiving device, the extended and largely unknowable communication network with its many powered interconnecting technologies? Though such conundrums may be approached as matters of accounting—ones laced with important questions of responsibility and governance, particularly when energy consumption is turned into carbon units (Barrett et al. 2013)—doing so maintains a focus upon metrics of energy demand rather than understandings of what energy is for.

To engage more with what energy is for, it is useful to ask not how we can locate energy use in such examples of complex practices, but how we can investigate the networks and various forms of relationality that social theorists of space prioritise in their accounts (Thrift 2006). Manifestly the social world is not just played out across a continuous physical surface, but is cut through with varied forms of interconnection between people, phenomena, ideas, ways of living, technologies and much else. These interconnections are enacted through the mobilities and flows of varied chains of people and things (Urry 2007), and the importance of these social phenomena has led to discussions not only of the compulsion to proximity (Boden and Molotch 1994) but also the existence of network capital which facilitates forms of physical or digital connectedness (Larsen et al. 2006). Within social scientific research, acknowledging and investigating connections, networks and proximity has thus helped to develop a language for the empirical analysis of varied topological relations and the stakes of their reproduction or change (e.g. Mol and Law 1994).

Building upon this work, it becomes clear that energy consumption takes place through various networks and forms of relationality. Rather than commenting upon Cartesian maps of airplane routes, consideration is instead focused on how a moving airplane enables a set of relations between points of departure and arrival that transcends their physical separation and distance. Similarly, a text message provides an instantaneous proximity and intimacy between distantly located people that serves to collapse geography as more conventionally understood. Energy demand is caught up in varied ways in such flows and interconnections—the flying of the airplane, the movement of the text message—and sustains and enables their reproduction. Moreover, expectations, norms and institutions become established around and in relation to these energy demanding flows; see, for example, in this volume Day et al. on expectations of long distance leisure travel, and Jones et al. on the changing norms of virtual and co-present collaboration in business. Energy demand and most cases of contemporary networks of flows are thus closely interconnected.

Taking a step further, it follows that space is not a given thing, but being continually produced as an outcome of social processes. Physical space evidently does exist and can be talked about in standard objective terms, but more integral to the processes and dynamics with which we are concerned in this book is socially produced space. As Lefebvre notes, "Every social space is the outcome of a process with many aspects and many contributing currents... In short, every social space has a history" (1991: 110). Approaching space in this way as "the product of interrelations" (Massey 2005) that are continually being made, and as the product of material and immaterial flows that shift, reform and transform over time (Sheller and Urry 2004), presents different questions and opportunities for understanding social processes and how energy demand is both constituted and implicated in the making of different spatialities.

Firstly, approaching space as continually made and remade in practice brings into question many dichotomies and apparently straightforward relations of relevance for understanding what energy is for. For example, the home as an intensively lived everyday space has been the focus of much critical examination engaging with the multiplicity of discourses and ideals with which it has historically become infused (Blunt and Dowling 2006). Beillan and Douzou, in their chapter in this volume, take on some of the classic binaries applied to the home (public-private, indoor-outdoor), showing empirically how the space of the home is produced through living in a shifting profile of materialities and meanings that are sustained, remade, acquired and discarded over time. Faced with such fluidity, applying binary spatial categories rapidly becomes problematic. For example, seeing the home as only 'indoor' is problematic, they argue, because for its occupants it is in continual interaction with 'outdoor' in terms of the use and meanings of windows, doors and balconies. As energy flows are also mediated by those relations (see also Hitchings 2011), this questioning of dichotomies uncovers important dynamics obscured by how spaces have been traditionally categorised. Other examples of problematic spatial categories feature across the contributions in this volume, including in the chapters by Burkinshaw and Mullen and Marsden, who discuss how the notion of distinct 'workplaces' and pathways of commuting to and from fixed locations is becoming increasingly incoherent with the realities of when and where work and employment is being enacted.

Secondly, investigating how spaces are made and remade highlights the extent to which demanding energy is an intrinsic part of such transformations. Cities are the most emblematic materialisations of spatial transformation and are routinely associated with a density of energy use, arising both from the intensity of urban activity and the assemblages of people, technologies, institutions, infrastructures and much else that make up the production and reproduction of contemporary urban space (Bulkeley et al. 2014; Rutherford and Coutard 2014). In spatial terms cities are where diverse networks of flow most intensely come together, necessarily dependent on energy both for the interrelations through which they are made as city spaces, and held together as (imperfectly) functioning and developing social, economic and infrastructural systems. Urban infrastructures support all sorts of activity and practice performances, evolving in their form and prevalence as patterns of infrastructural relations and interconnections shift over time (Shove et al. 2015); a co-evolving set of processes captured to some degree in Wiig's chapter in this volume focused on digital connectivity.

Whilst the energy dependence of such intensely realised space-making is apparent in general terms, in Allen's chapter in this volume, this becomes far more explicit. Following how rural green fields are dramatically transformed for short periods of time through the making of a music festival, he shows how the coming together of multiple co-terminus flows (of people, vehicles, performers, toilets etc.) actively works in combination to turn rural space into a sort of temporary urban one: an appropriate place for sleeping, cooking, eating, taking drugs and dancing, rather than for solitary rambling or animal grazing. While such a special case of spatial transformation is dramatic and striking—bringing energy infrastructure and energy use into an explicitly new (if temporary) set of spatial relations—it is also revealing of the slower processes of transformation generally involved in space-making and remaking, as well as of the established relations that are temporarily 'left behind' when everyday practices are relocated and performed to some degree differently—as they are not only in festival-going, but in many other instances of living elsewhere, holidaying or visiting (see examples of such instances in chapters by Sahakian and Day et al. in this volume).

Thirdly, the unevenness in the processes through which space is made and remade has implications for how demanding energy is also differentially constituted. For many engaging with the spatial in relational terms, power and politics are central concerns, drawing out both how past and current power relations are reflected in the spatial configurations that now exist, but also the possibility for these to be made differently in the future. Massey (2005: 85), for example, argues that the "intrinsic relationality of the spatial, is not just a matter of lines on a map: it is a cartography of power". While none of the contributions to this volume directly conceptualise their analyses of demanding energy in terms of uneven power relations, we can certainly find these lurking in the shadows. In a focus strikingly different from that of much of the existing energy and inequality literature concerned with the fuel poor (e.g. Harrison and Popke 2011; Chard and Walker 2016), Sahakian shows how wealthy expat households in Switzerland have a particular agency in space-making and in the production of energy demand. They are able to fill their voluminous homes with multiple, large, expensive energy-using appliances and employ staff to keep them clean and tidy at all times in order to (re)produce a status expressed through the qualities of where and how they live. Their self-acknowledged privilege, located in a key geographical node in the global circulation of finance and capital, is thus intrinsic to the materiality of their distinctive domestic spaces, with demanding energy a necessary and substantial ingredient.

In stark contrast, Mullen and Marsden make clear how the households in their study are caught in very different cartographies of power, with instabilities and uncertainties in the network of spatial relations interconnecting home, work and education that make car dependence a hard to avoid and afford necessity. The need to travel from home to sites of work or education across physical space, and to use the energy-intensive car to achieve this is, in their account, overlain with economic and social relations formed by such things as zero-hour or temporary contracts, problematic landlords, mortgage payments, bus routes and the health of family members. For these households the social production of space therefore brings fragility to their performed, routine days and problematic consequences for well-being.

In such ways then we can see that taking on space in a more sophisticated way, recognising its multiplicity, relationality and dynamism, has much to bring to understandings of what energy is for. Whilst not often a dominant or overt element in spatial narratives, energy demand is integral to the making of past, present and future socio-spatial relations (Calvert 2016), an observation pertinent not only to the empirical sites encompassed in contributions to this volume, but also to the production of space in many other contexts, settings and parts of the world we have not been able to extend to.

1.2 Time and Demanding Energy

Just as energy use takes place in spaces and through spatial relations, so too it occurs in time. Time can also be considered as both independent from and intertwined with human activities. Early social theorists of time made a distinction between natural time and social time (Sorokin and Merton 1937; Evans-Pritchard 1939). Natural time is the temporality of natural processes—astronomical cycles, ecological rhythms, seasons, tides—while social time is a construct by which we organise, and experience, everyday life. The two are linked, in that social time is to some extent structured by the natural rhythms of day and night, annual cycles and so on. As humans we are also subject to our own 'natural' bodily processes which have cyclical and linear temporalities, such as bodily ageing. Both, moreover, have implications for what energy is for. Despite technological developments, the temporalities of nature that underpin social organisation continue to have clear bearing on energy demand (Walker 2016): fairly obviously, we use lighting in hours of darkness, heating in colder seasons, cooling in warmer ones. Apart from such relatively basic needs, other social conventions are pinned to natural cycles and may occasion energy demand, for example for mobility for summer holidays, or lighting for winter festivals.

Although social time may have some anchoring in natural or ecological time, (and leaving aside for now reflections on how we have altered 'natural' processes and their temporality), like socially produced space, it is a more constructed and multifaceted phenomenon. Social time arguably became more decoupled from ecological time with the industrial revolution and the widespread deployment of clocks. While time is at its basis a relational quality, a means of orientation or ordering, clock time is a social institution that quantified and measured the abstract, and created a system of accounting and control that was fundamental to the development of capitalism (Adam 1990, 1995; Thompson 1967). Clock time facilitates temporal processes such as scheduling and coordination, in terms of the temporal placement and duration of activities. Through such means, it enabled mass mobility, mass production, education, shared leisure and many aspects of consumption.

The social production of clock time has many facets and implications for energy demand. For one, we often live by schedules, and as our activities are scheduled, so is our energy demand (Walker 2014). Coordinating practices around normal working hours, for example, contributes to regular peaks in demand: of traffic during twice daily commuting 'rush' hours and of domestic electricity and gas consumption during evening periods of cooking, eating and leisure. These peaks are increasingly problematic for energy production and associated emissions, as roads and power stations approach or exceed capacity during these times. As a result, they need to be planned for, and capacity, which is often redundancy, to be built in (Torriti et al. 2015; Torriti 2016). Temporal peaks of energy demand are therefore phenomena that lead to further consumption of resources, time (as in time spent in traffic congestion) and energy itself—see for example Wiig, this volume, on the energy consumed by data centres awaiting peak internet traffic.

Given the problematic peaking of energy and mobility demand, there is growing interest in flexibility: the potential to produce time differently through the de-synchronisation of activities at a societal level so that they are spread more evenly through the day, week or year. We can see such a move in German and Dutch policies of scheduling school holidays over different weeks in different regions of the country to reduce pressure on transport systems and leisure facilities. Investigating how rhythms and synchronisation are enacted within institutions is an important line of investigation, as shown in the contributions to this volume by Curtis et al. and Blue, in order to understand how flexibility might contribute to different temporal arrangements.

As our authors highlight, however, it is easier to reconfigure the temporal relations of some practices than others. Durand-Daubin and Anderson's chapter, for example, reveals the perhaps surprising temporal obduracy of French lunching and dining routines, and how such obduracy is less apparent in the UK. They suggest that one key to understanding this may be to investigate what activities cooking and eating are sequenced with—an interrogation of temporal sequences at a daily level thus informing the analysis of change over longer time periods. The temporal relations around travel can similarly be challenging to reconfigure. Burkinshaw's chapter, which discusses the limits to the take-up of flexible working hours, finds that a major reason for the continuation of rush hour journeys is the temporal sequencing of journeys to work with other necessary household activities, notably journeys to school. Problems can arise then through lack of coordination, or when flexibility in one set of practices or activities meets inflexibility in another. Southerton (2012) identifies people's difficulties in coordinating practices in time as underlying generalised feelings of harriedness and time-related anxiety. Energy demand may also be implicated: Mullen and Marsden (this volume) identify unpredictable flexibility in working hours as resulting in more resource intensive means of travel as people opt to keep a car, because inflexible—and at the same time sometimes unreliable—transport schedules cannot meet their needs. These contributions highlight that although social time is constructed through activity, the extent to which temporal relations can be remade is an important focus for research on demanding energy.

Whilst on one hand the units of clock time facilitate discussions of particular schedules, peaks, synchronicities and opportunities for flexibility, this ability to count and account for time also tends to make us think of time as a resource. Time can appear to be something that is consumed by practices (Shove 2009). We only have so much time, and practices use it up. Practices also compete for our time. For example, in Day et al.'s contribution to this volume we see how leisure travel among retired people rivals for time with family responsibilities and caring practices, while in Durand-Daubin and Anderson's chapter, cooking and eating are seen competing with work as work hours become more flexible. The effects on energy demand are complicated and hard to predict; in some instances it may mean less of an energy consuming practice such as long distance leisure travel, or cooking, but it could mean more shifting of energy consumption to other spaces such as restaurants and take away outlets.

Thinking of time as a resource, and one often with monetary value (Adam 1990), also leads to perceptions of wasted time, and the need to save time. Saving time can increase energy consumption as we enlist appliances to help us to perform tasks faster, or remotely: see Greene, this volume. The notion of time-as-resource might also lead us to think about whose time matters and whose time is used for what. Some people such as domestic staff in Sahakian's chapter, or indeed some of the husbands in Greene's, are engaging in energy consuming practices so that other people's time can be freed up. Our outsourcing of energy consuming practices, such as cooking, is as much about time budgeting as anything. And, as Mullen and Marsden's chapter illustrates, one person's flexibility—such as an employer's—is another person's chaos and wasted time as they wait for public transport at badly served hours.

Approaching time as something that practices consume prompts valuable explorations into processes of demanding energy, but as Shove (2009) suggests, we might also, alternatively, see practices as producing time. That is, the temporality of our lives is emergent from the nature and rhythms of the practices that we engage in. Quantitatively, we might gauge the passing of time by the completion of particular activities, projects or life stages. Qualitatively, periods of time take on a particular significance or character because of the practices that we engage in—for example, the holidays are the holidays because we do certain things and don't do certain others.

In considering the qualitative aspects of time, Cipriani (2013) discusses a distinction in classical Greek between *chronos*, understood as sequential, linear or cyclical time, and *kairos*, signifying the right or proper time for something (see also Szerszynski 2002). Kairological time is highly relevant for understanding what energy is for, as it relates to the organisation of routines and the extent to which temporal flexibility is possible. Psarikidou, in this volume, discusses kairological time in notions of when it is safe or not safe to take public transport, in that case linked also to natural rhythms of day and night. In Day et al.'s chapter, taking a longer perspective, retirement is seen in a kairological sense by their research participants as a time of freedom and the time of life for leisure travel. Given the existence of quite energy intensive travel such as cruising, this has significant implications for the patterning of energy consumption over the life course, and also in society more broadly, as demographic patterns shift.

As this brief discussion has illustrated, theoretical concepts related to dynamics of social time are important resources for describing patterns within social practices and energy consumption, as well as for developing rich discussions of how social practices are constituted and changing. Whilst an interest in change has been implicit in our discussions of both space and time, the next section turns to address this more explicitly.

1.3   Change and Demanding Energy

The importance of change to understanding energy consumption might, on the face of it, seem obvious. Existing discussions of energy demand, after all, place considerable emphasis upon making certain changes in order to do things such as optimise systems or manage and reduce overall consumption. Though these are important concerns, and the authors in this collection seek to contribute in various ways, change is not primarily discussed in this more instrumental sense. The same move from nouns to verbs that leads us to focus upon demanding energy rather than energy demand as a thing applies here—authors are more concerned with studying the changing dynamics of what energy is for than particular changes that will impact energy demand.

This approach can be tied back theoretically to the understanding that the social world is constituted by practices. The model of agency underpinning this idea, as expressed in theories of practice, is one built upon the principle of indeterminacy. As Giddens suggests, "it is a necessary feature of action that, at any point in time, the agent 'could have acted otherwise'" (1979: 56). Schatzki concurs that people's activity is in this way fundamentally open, regardless of context: "No matter how strongly his or her ends, desires, hopes, preferences, and the like 'point toward,' or even 'single out,' a given path of action, nothing guarantees that it or any other particular action is performed" (2002: 232). Seeing action as indeterminate thus becomes an important prompt for its empirical study. Processes and practices must be interrogated because their activities cannot be taken for granted. Further, and of particular importance for studies of demanding energy, Giddens notes that this study of the reproduction of practices must come before considering any consistency in their outcomes or consequences (1979: 214). That particular practices demand more or less energy than others, for example, should be considered only after having grasped how the practices themselves are reproduced.

A question then arises as to whether focusing on practices privileges the reproduction of practices over their transformation or change. Some descriptions of practices might appear to privilege stability in this way—Reckwitz for example makes the very strong claim that: "For practice theory, the nature of social structure consists in routinization. Social practices are routines" (2002: 255). He follows on to suggest that change must then occur through "everyday crises of routines" (2002: 255). Whilst identifying particular practices as units of enquiry requires some understanding that they consist of similarly patterned activities that are repeatedly performed, this foregrounding of routines is unhelpful because of how it discourages recognition of the indeterminacy of action. That is, misunderstandings can easily arise that routine is more foundational to or inevitable within practices than change. At other times, authors may appear to emphasise change—as in Giddens' statement that "Change, or its potentiality, is thus inherent in all moments of social reproduction" (1979: 114). Whilst this is consistent with the idea of indeterminacy, and with the acknowledgement that "Absolute repetition is only a fiction of logical and mathematical thought" (Lefebvre 2004: 7), it could also lead to misunderstandings that all action is change. Ultimately, theories of practice do not prima facie privilege either routine and stability or trans-formation and change—"stability and change come together in the social site" and researchers are the ones who identify differences of consequence (Schatzki 2002: 254).

For the authors in this collection, studying changing practices is thus in part about identifying differences of consequence for understanding what energy is for. In many instances, these consequential (though some-times unintended) differences are inextricable from space and time, which come together in societal schedules and cycles as well as more sig-nificant social transformations that reconfigure spatio-temporal relations (Giddens 1979: 205). As societal conventions, tastes, policies, institu-tional organisations, population dynamics and economic relations evolve through practices we can identify changing temporal patterns of energy demand over the long term. In Greene's chapter, for example, we see how institutional change in Ireland affecting policies and norms around wom-en's work impacted the distributional energy intensity of domestic prac-tices in individual households. We see similar longer-term change in norms and expectations in Day et al. and Durand-Daubin and Anderson's chapters.

Such change in temporal patterns of practices and associated energy demand is inextricable from spatial transformations—especially related to the evolving affordances of technologies and infrastructures (Shove 2009). The growth in aviation over recent decades enables routine long haul travel of Day et al.'s retirees and Jones et al.'s global consultancy employees, among others. Internet-enabled technologies and data-related infrastructure, as discussed by Wiig, allow practices to be bundled and layered—travel can be performed at the same time as work, learning or social interaction, increasing energy consumption on the one hand, but potentially also changing feelings towards public transport as time spent on it is no longer 'wasted', as Psarikidou discusses. Such technologies have also enabled the at least partial de-synchronisation of work practices, though considerable pull still exists to co-presence and synchronisation, as Burkinshaw discusses.

The shape and trajectories of longer-term transformations are also undoubtedly affected by the particular sites under consideration. As Giddens argues: "all social change is conjunctural. That is to say, it depends upon conjunctions of circumstances and events that may differ in nature according to variations of context" (1984: 245). Focusing as it does upon cases from developed countries in the Global North, we therefore acknowledge that this book is limited in its purview. After all, if action is inherently indeterminate, then it is important to study a wide range of practices, in diverse spaces and times, in order to ensure that inappropriate summaries are not made. Therefore whilst we begin to develop a set of concepts, methods and cases appropriate for the study of demanding energy in this book, it will be important that they are further tested and developed in dialogue with cases from other regions and social, political and economic sites in the future.

The final quality of change that is central to the chapters of this book, alongside indeterminacy and spatio-temporal transformation, is interdependency. Whilst the actions contributing to any one practice are indeterminate, it is not enough to consider change in relation to single practices alone. Demanding energy therefore must be interrogated in relation to what Giddens, quoting Etzioni, suggests is a fundamental "interdependence of action: in other words, to 'a relationship in which changes in one or more component parts initiate changes in other component parts, and these changes, in turn, produce changes in the parts in which the original changes occurred'" (Etzioni 1968 in Giddens 1979: 73). The multiple units that can become embroiled in demanding energy—people, practices, technologies, meanings and institutions—are thus shown to transform in relation to each other in expected and unexpected ways. This is important for understanding existing and potential flexibility in practices and energy demand. The chapters by Blue and Curtis et al. take up this concern within different institutional contexts. For Blue, the importance of finding ways to change energy demand within hospitals is temporarily set aside in order to better understand how hospital practices are reproduced and how changing connections between working units, often facilitated by new technologies, affect rhythms of consequence for treatment scheduling, duration of technology use and provision of care. Curtis et al. look at hotels, examining how demand management processes involve not only particular technologies but also a range of interdependent human activities. Here the potential for flexibility in demand is carefully linked to networked technologies, people, and hotel practices and events.

By attending to specific cases, the chapters of this book describe how particular relations are changing, and then establish important insights from this about how processes of demanding energy have and might be changing. The final section of this introduction outlines how these contributions are organised.

1.4 The Book Structure

As this collection represents the first sustained effort to develop an agenda for studying the social dynamics of energy demand, its aim is to inspire new ways of thinking and working as much as to consolidate a body of new evidence or set of relevant theoretical concepts. The structure of the book, as a result, is not driven by particular categorisations of space, time and change, or by familiar distinctions (e.g. between work, home and transport practices). Undoubtedly other connections exist amongst these pages—including ones that focus upon familiar categories of working (Burkinshaw, Jones et al., Blue), transport (Psarikidou, Day et al., Burkinshaw, Mullen and Marsden), material infrastructures (Wiig, Allen, Curtis et al.) or domestic life (Douzou and Beillan, Greene, Sahakian, Durand-Daubin and Anderson). We have, however, curated sets of chapters that have the potential to start useful conversations about how multiple studies and approaches to demanding energy might fit together to form a broad basis of evidence informing future work and policy. The pairs or trios of chapters in some instances address quite different empirical cases and types of data, or diverge in terms of engagement with policy, industry or theoretical concerns. In this way, they push against established boundaries and vocabularies within energy demand research. This friction is strategic and consistent with our aim of establishing different vocabularies, methodologies and evidence for understanding processes of demanding energy.

The parts of the book highlight six themes of key importance for processes of demanding energy: Making connections, Unpacking meanings, Situating agency, Tracing trajectories, Shifting rhythms and Researching demand. Before each part is an 'interlude', which outlines the chapters in more detail, noting where they can be seen to build upon similar approaches, where their analyses offer different but complementary insights and where they come together to contribute to understandings of how energy demand is constituted, patterned and changing.

In the first part, *Making connections*, authors explore the systemic interdependencies that contribute to the constitution and change of energy demand. Material and infrastructural relations play a significant role in the discussions and authors raise important questions about how these are connected to varied practices, corporate development plans, group norms, conveniences, and comparisons made between both exceptional and mundane events.

The second part, *Unpacking meanings*, discusses dynamics of space, time and change from the perspective of those (re)producing them. The chapters bring into view the experiences and meanings of those engaged in everyday travel and inhabiting homes, and demonstrate how these exceed and challenge established categories within energy demand discourses.

The third part, *Situating agency*, discusses how the interplay between structure and agency can be considered in relation to the interdependence of different practices, approaching agency as enacted in relation to specific spatial and temporal constraints and practice sequences. Each of the chapters focused on accounts of negotiating working, commuting, caring and travelling, among other things, demonstrate the limitations of assumptions routinely made about individual capacities to act differently.

The fourth part, *Tracing trajectories*, takes a long temporal view, with chapters tracing particular processes of longer-term change in cooking and eating, domestic work and business travel. The shifting dimensions of working practices particularly figure in this part of the book, with the chapters each showing how the implications of changes in both who is doing work and how work is organised and carried out have a host of related consequences, including directly and indirectly for patterns of energy demand.

The fifth part, *Shifting rhythms*, focuses on the temporal orderings that make up the everyday, and that give rhythm to social processes and in turn to how energy demand fluxes and flows over time. The chapters here are concerned with how temporalities are grounded within particular settings and situations (hotels and hospitals), taking different approaches to understanding how much purposeful flexing to rhythms of energy use can really be achieved in order to fit with wider energy system objectives.

The book concludes in the sixth part, *Researching demand*, with a discussion of research design and methodology, reflecting back on all of the chapters to consider the ways in which they have articulated and framed their research questions and how issues related to the selection of cases and samples have been addressed. In so doing the intention is to provide a discussion and an inspiration that can be taken forward into future research that seeks to understand more about what energy is for and the social processes through which demanding energy is enacted.

Bibliography

Adam, B. 1990. *Time and social theory*. Cambridge: Polity Press.

———. 1995. *Timewatch: The social analysis of time*. Cambridge: Polity Press.

Barrett, J.R., G. Peters, T. Weidmann, et al. 2013. Consumption-based GHG emissions accounting in climate policy: A UK case study. *Climate Policy* 13: 451–470.

Blue, S., and N. Spurling. 2017. Qualities of connective tissue in hospital life: How complexes of practices change. In *The nexus of practices: Connections, constellations, practitioners*, ed. A. Hui, T.R. Schatzki, and E. Shove, 24–37. London: Routledge.

Blunt, A., and R. Dowling. 2006. *Home*. London: Routledge.

Boden, D., and H. Molotch. 1994. The compulsion to proximity. In *Nowhere: Space, time, and modernity*, ed. R. Friedland and D. Boden, 257–286. Berkeley: University of California Press.

Bulkeley, H., V.C. Broto, and A. Maassen. 2014. Low-carbon transitions and the reconfiguration of urban infrastructure. *Urban Studies* 51: 1471–1486.

Burholt, V., and G. Windle. 2006. Keeping warm? Self-reported housing and home energy efficiency factors impacting on older people heating homes in north wales. *Energy Policy* 34: 1198–1208.

Calvert, K. 2016. From 'energy geography' to 'energy geographies': Perspectives on a fertile academic borderland. *Energy Policy* 40: 105–215.

Chard, R., and G. Walker. 2016. Living with fuel poverty in older age: Coping strategies and their problematic implications. *Energy Research and Social Science* 18: 62–70.

Cipriani, R. 2013. The many faces of social time: A sociological approach. *Time & Society* 22: 5–30.

Department for Transport. 2015. *Understanding the drivers of road travel: Current trends in and factors behind roads use.* London: Department for Transport. Available at https://www.gov.uk/government/uploads/system/uploads/attachment_data/file/395722/understanding-the-drivers-road_travel.pdf

Etzioni, A. 1968. *The active society.* New York: Free Press.

Evans-Pritchard, E.E. 1939. Nuer time-reckoning. *Africa* 12: 189–216.

Giddens, A. 1979. *Central problems in social theory.* London: Macmillan Press.

———. 1984. *The constitution of society.* Cambridge: Polity Press.

Harrison, C., and J. Popke. 2011. "Because you got to have heat": The networked assemblage of energy poverty in eastern North Carolina. *Annals of the Association of American Geographers* 101: 949–961.

Hitchings, R. 2011. Coping with the immediate experience of climate: Regional variations and indoor trajectories. *WIREs Climate Change* 2: 170–184.

Hui, A. 2017. Variation and the intersection of practices. In *The nexus of practices: Connections, constellations, practitioners*, eds. A. Hui, T.R. Schatzki, and E. Shove, 52–67. London: Routledge.

Hui, A., T.R. Schatzki, and E. Shove, eds. 2017. *The nexus of practices: Connections, constellations, practitioners.* London: Routledge.

Inderwildi, O., and S.D. King, eds. 2012. *Energy, transport, & the environment: Addressing the sustainable mobility paradigm.* London: Springer-Verlag.

Isaac, M., and D.P. van Vuuren. 2009. Modeling global residential sector energy demand for heating and air conditioning in the context of climate change. *Energy Policy* 37: 507–521.

Larsen, J., J. Urry, and K. Axhausen. 2006. *Mobilities, networks, geographies.* Aldershot: Ashgate.

Lefebvre, H. 1991. *The production of space.* Oxford: Blackwell.

————. 2004. *Rhythmanalysis: Space, time and everyday life.* London: Continuum.

Lin, W. 2016. Re-assembling (aero)mobilities: Perspectives beyond the west. *Mobilities* 11: 49–65.

Lord, C., M. Hazas, A.K. Clear, et al. 2015. Demand in my pocket: Mobile devices and the data connectivity marshalled in support of everyday practice. In *Proceedings of 33rd Annual ACM Conference on Human Factors in Computing,* eds. B. Begole, J. Kim, K. Inkpen, et al., 2729–2738. Toronto: ACM.

Massey, D. 2005. *For space.* London: Sage.

Mokhtarian, P.L., and I. Salomon. 2001. How derived is the demand for travel? Some conceptual and measurement considerations. *Transportation Research Part A: Policy and Practice* 35: 695–719.

Mol, A., and J. Law. 1994. Regions, networks and fluids: Anaemia and social topology. *Social Studies of Science* 24: 641–671.

Morley, J. 2017. Technologies within and beyond practices. In *The nexus of practices: Connections, constellations, practitioners,* ed. A. Hui, T.R. Schatzki, and E. Shove, 81–97. London: Routledge.

O'Doherty, J., S. Lyons, and R.S.J. Tol. 2008. Energy-using appliances and energy-saving features: Determinants of ownership in Ireland. *Applied Energy* 85: 650–662.

Reckwitz, A. 2002. Toward a theory of social practices: A development in culturalist theorizing. *European Journal of Social Theory* 5: 243–263.

Rinkinen, J. 2015. Demanding energy in everyday life: Insights from wood heating into theories of social practice. PhD Essay dissertation, Aalto University.

Rutherford, J., and O. Coutard. 2014. Urban energy transitions: Places, processes and politics of socio-technical change. *Urban Studies* 51: 1353–1377.

Sahakian, M.D. 2011. Understanding household energy consumption patterns: When "west is best" in metro manila. *Energy Policy* 39: 596–602.

Schatzki, T.R. 1996. *Social practices: A Wittgensteinian approach to human activity and the social.* New York: Cambridge University Press.

————. 2002. *The site of the social: A philosophical account of the constitution of social life and change.* University Park: Pennsylvania State University Press.

Sheller, M., and J. Urry. 2004. Places to play, places in play. In *Tourism mobilities: Places to play, places in play,* eds. M. Sheller and J. Urry, 1–10. New York: Routledge.

Shove, E. 2009. Everyday practice and the production and consumption of time. In *Time, consumption and everyday life: Practice, materiality and culture,* eds. E. Shove, F. Trentmann, and R. Wilk, 17–33. Oxford: Berg.

————. 2010. Beyond the ABC: Climate change policy and theories of social change. *Environment and Planning A* 42: 1273–1285.

————. 2017. Matters of practice. In *The nexus of practices: Connections, constellations, practitioners*, eds. A. Hui, T.R. Schatzki, and E. Shove, 155–168. London: Routledge.

Shove, E., and G. Walker. 2014. What is energy for? Social practice and energy demand. *Theory, culture and society* 31: 41–58.

Shove, E., M. Pantzar, and M. Watson. 2012. *The dynamics of social practice: Everyday life and how it changes*. London: Sage.

Shove, E., M. Watson, and N. Spurling. 2015. Conceptualising connections: Energy demand, infrastructures and social practices. *European Journal of Social Theory* 18: 274–287.

Sorokin, P.A., and R.K. Merton. 1937. Social time: A methodological and functional analysis. *American Journal of Sociology* 42: 615–629.

Southerton, D. 2012. Habits, routines and temporalities of consumption: From individual behaviours to the reproduction of everyday practices. *Time & Society* 22: 335–355.

Szerszynski, B. 2002. Wild times and domesticated times: The temporalities of environmental lifestyles and politics. *Landscape and Urban Planning* 61: 181–191.

Thompson, E.P. 1967. Time, work-discipline and industrial capitalism. *Past & Present* 38: 56–97.

Thrift, N. 2006. Space. *Theory Culture and Society* 23: 139–155.

Torriti, J. 2016. *Peak energy demand and demand side response*. London: Routledge.

Torriti, J., A. Druckman, B. Anderson, et al. 2015. Peak residential electricity demand and social practices: Deriving flexibility and greenhouse gas intensities from time use and locational data. *Indoor and Built Environment* 24: 891–912.

Urry, J. 2007. *Mobilities*. Cambridge: Polity Press.

Walker, G. 2014. The dynamics of energy demand: Change, rhythm and synchronicity. *Energy Research & Social Science* 1: 49–55.

————. 2016. Rhythm, nature and the dynamics of energy demand. In *Proceedings of DEMAND conference*, Lancaster. Available at http://www.demand.ac.uk/wp-content/uploads/2016/03/DEMAND2016_Full_paper_57-Preston.pdf

Watson, M. 2017. Placing power in practice theory. In *The nexus of practices: Connections, constellations, practitioners*, eds. A. Hui, T.R. Schatzki, and E. Shove, 169–182. London: Routledge.

Wiig, A. 2013. Everyday landmarks of networked urbanism: Cellular antenna sites and the infrastructure of mobile communication in Philadelphia. *Journal of Urban Technology* 20: 21–37.

Williams, J. 2012. *Zero carbon homes: A road map*. Abingdon: Routledge.

Allison Hui is an Academic Fellow in Sociology at the DEMAND (Dynamics of Energy, Mobility and Demand) Research Centre at Lancaster University, UK. Her research examines transformations in everyday life in the context of changing global mobilities, focusing particularly on theorising social practices, consumption and travel. Her recent co-edited publications include *The Nexus of Practices* (Routledge, 2017) and *Traces of a Mobile Field* (Routledge, 2017).

Rosie Day is a Senior Lecturer in Human Geography at the University of Birmingham, UK. Her research interests centre around social inequalities in access to and experience of environmental and energy resources, with a related interest in environments of ageing and older age. She works collaboratively on a number of multidisciplinary projects in diverse international contexts.

Gordon Walker is Co-Director of the DEMAND (Dynamics of Energy, Mobility and Demand) Research Centre at Lancaster University and Professor in the Lancaster Environment Centre. He has wide ranging expertise on the social and spatial dimensions of sustainable energy issues, sustainable social practices and cross-cutting issues and theories of climate, energy and environmental justice.

Part 1

Making Connections

Once again, I have lost my mobile phone cable. No matter how organised I try to be, it is never where I expect. Very unlike charging places. They seem to be everywhere. At my work computer. The power outlet beside my train seat. The USB port built into my fancy new sofa. But what good is a charging outlet if you can't find the cable? It's not like I could just leave my phone battery dead either. People expect me to be connected. To be in contact at all times in case my son's cough has gotten worse at preschool, or the delivery person gets lost trying to deliver my package. I've come to rely on it so much. No way could I go back to having my personal and work calendars on paper. But maybe I'll find out what disconnection is like if I can't find the stupid cable.

As the first three chapters of this book demonstrate, understanding how energy demand is constituted and transforming depends upon tracing a diverse set of connections between people's practices and the energy-using technologies, infrastructures and social norms upon which they depend. Some of these connections are materialised quite obviously within everyday spheres—mobile phone cables are tangible evidence that this technology and the multiple practices it facilitates intrinsically depend upon energy in the form of electricity. Yet as these chapters highlight, energy demand must also be understood as embedded systemically, in ways that stretch beyond the activity or observation of any one person. The use of mobile phones is thus connected to the practices of companies that develop and provide both electrical and digital communications infrastructures, which support the growing assumption that mobile phones should be used everywhere. Social practices, infrastructures and the shared understandings that underpin and reinforce both are co-produced, with important consequences for energy demand.

In different ways, all three chapters are interested in what can be learned from the transformation and normalisation of particular infra-structural arrangements that facilitate patterns of energy demand and provide part of the setting amidst which everyday practices occur. Whether considering the evolution and lock-in of material arrange-ments within the home, the embedding of new infrastructures to sup-port digital connection on the move, or the dynamics of temporary infrastructures supporting music festivals, the authors demonstrate how 'normal' levels of energy demand for particular practices depend upon infrastructures. What exactly 'normal' is, however, both varies across space and is continually changing. Whilst both festival goers and wealthy expats cook meals and wash themselves, they do so with differ-ent understandings of what is 'good enough', amidst very different infrastructural arrangements and using differing levels and forms of energy. In both cases there is also an identifiable historical transforma-tion wherein the energy routinely demanded by these practices in these settings has been steadily increasing, alongside changing standards for cooking and washing. The chapters thus take infrastructures to be dynamic, and multiply connected to everyday practices, as studies of science and technology have well established. As a consequence, energy demand is constituted in part through the co-production of infrastruc-tures, which are situated at the intersection of the practices of those who design, assemble, provision, manage and use them.

Taking different spatio-temporal framings, each author highlights often overlooked aspects of such intersections. Alan Wiig juxtaposes a fictionalised train journey with a story of proposed changes to make digital infrastructures more robust, enabling mobile communications to expand and take on even more territory. Whilst these two stories pivot around the same infrastructures, the actors and materialities that are central to one are largely absent in the other. Marlyne Sahakian focuses on how practices and materials are normally arranged within the home. Yet the 'normal' arrangements under investigation are unlike those many people experience, as they facilitate the lives of wealthy expats for whom, for example, owning and using multiple dishwashers is a part of ordinary life. The question thus becomes how historically and comparatively excessive infrastructures are normalised within a particular context. Michael Allen's chapter takes on an explicit temporal framing, asking how 'temporary' events are dependent upon social and infrastructural dynamics that are in fact anything but fleeting. In effect, his chapter asks what happens if the seemingly exceptional and excessive demand associated with temporary events and infrastructures are seen in a different light. He shows that moving beyond duration to think about other temporalities and spatialities of events raises important questions about how the energy demand of music festivals compares with the patterns of everyday living that they temporarily displace.

As each chapter illustrates, the spatio-temporal dynamics of these processes are crucial to understanding how particular connections matter for energy demand. The specific siting of infrastructures is highly consequential, whether in terms of shared social understandings of 'normal' home appliances, the extent to which the provision of mobility and digital communications infrastructures can overlap, or the need for temporary, rather than more permanent, installations. So too are historical developments that can establish implicit understandings surrounding need, excess and accessibility, and lead to the ratcheting up of energy demand. Most importantly, however, the authors raise important questions about the potential for change. While increasing the accessibility of digital infrastructures may promote convenience, it also comes with consequences for overall levels of energy demand and the potential for even more data centres that use energy to keep the possibility of digital connection always available. While household appliances and professional cleaners who help to use them might be convenient for sustaining a presentable home, their presence can also lead to situations where cleaning is done out of habit or contract rather than 'necessity'. While music festivals are easily singled out as requiring problematic levels of energy demand, pointing to how they also involve low-energy forms of everyday living raises important questions about which energy-using practices need to change. Whilst there is no straightforward answer to how practices, infra-structures and social norms together contribute to the constitution of energy demand, these chapters point to many of the connections that need to be better understood in order to start thinking creatively about future transformations.

2

Demanding Connectivity, Demanding Charging: The Co-production of Mobile Communication Between Electrical and Digital Infrastructures

Alan Wiig

2.1 Introduction: Charging Smartphone Batteries, Powering the Internet

For many if not most people, the experience of travel goes metaphorically and literally hand-in-hand with a constant connection to a smartphone or other mobile computing device. Over the course of less than a decade,[1] these devices and emergent practices have transformed the social utility of space. Proximity and distance are re-formatted through wireless digital connectivity; a sidewalk or a train car becomes a space for a (not-so) private conversation or, through a laptop and its software, an extension of an office. Social media, the general Internet, locative media and mapping applications are all accessed through these pocketable computers. All these uses demand energy, in the form of electricity, and not solely through maintaining a charge of the device's battery. Every stage of communication from connection to transmission, to storage and maintenance

A. Wiig (✉)
Urban Planning and Community Development, University of Massachusetts, Boston, MA, USA

© The Author(s) 2018
A. Hui et al. (eds.), *Demanding Energy*, DOI 10.1007/978-3-319-61991-0_2

of digital information requires electricity. As use of smartphones becomes prevalent throughout the day and across multiple practices, from home to work and in transit between, the need to maintain connection to the pervasive, wireless Internet, and to maintain a device's charge away from the readily accessed electricity at the home and at work means multiple infrastructures, social practices, and business opportunities are arising around patterns of mobility, smartphone use, and energy provision.

By number, mobile phone subscriptions are nearly equivalent to worldwide population; each device consequently requiring regular inputs of electrical energy (ITU 2014; Horta et al. 2016: 15). Sixty-eight percent of adults in the United States have an Internet-enabled smartphone and 45% have a portable tablet computer such as an Apple iPad (Anderson 2015). All these devices function through battery power; all these devices therefore must be charged regularly to remain useful. While the focus of these devices is on their wireless, connective abilities, this disposition functions through electricity.

The goal of this chapter is to conceptualize wireless digital connectivity as a matter of infrastructural co-production between electrical and digital communication infrastructure, and the energy demand of these systems and the social practices they support. Commonly termed mobile communication, this connectivity is a medium of social exchange predicated through telecommunication infrastructures. This connectivity is inseparable from the energy systems underlying *both* the telecommunications infrastructures and the charging provision powering smartphones and other personal computers. Building on Jasanoff's scholarship on co-production as "the constant intertwining of the cognitive, the material, the social and the normative" (Jasanoff 2004, 2005), an approach to co-production considers mobile, digital connectivity as an emergent social practice that necessitates also conceptualizing the energy demand of smartphones, the energy systems that power the devices, and the energy consumption of data centers holding the informational back-end enabling digital connectivity. Behind the expectation of connectivity lies multiple energy demands. The co-production of social exchange (communication, connectivity, and mobility) is constituted through layered infrastructures (transportation, telecommunication, and energy). Untangling energy demand thus entails both conceptualizing social practice around charging as well as the electrical demand of digital infrastructure itself. This chapter is not intended to provide a comprehensive overview quantifying the energy demand of these intertwined systems in any one location; the objective instead is to make visible emergent practices and, to quote Schatzki (2010), to consider these infrastructures as "material arrangements" facilitating charging in motion.

2.2 Background: The Energy Infrastructures of Mobile, Digital Connectivity

Pervasive, wireless digital connectivity is constituted through telecommunications and energy infrastructures, what Schatzki (2010) terms "material arrangements." Building off Schatzki to consider the infrastructures underlying energy demand, Shove et al. write that these arrangements "reflect and shape multiple social practices [...] and are, in turn, shaped by past and present forms of planning and design" (Shove et al. 2015: 1). In this approach, energy "is best understood as part of the ongoing reproduction and transformation of society", underlying the sites of social practice and "realized through artefacts and infrastructures that constitute and that are in turn woven into bundles and complexes of social practice" (Shove et al. 2014: 52). Digital, mobile connectivity as a social practice is constantly produced through its underlying demands for devices' electrical energy as well as the electricity and information-communication technology (ICT) infrastructures that provide the connection.

Briefly, to produce the constant wireless connectivity of a smartphone requires the ongoing broadcast of blanketing electromagnetic waves through wireless fidelity (WiFi) routers that can cover a small area such as a home, office, or cafe, with fast upload and download speeds, and cellular antennas that, mounted on tall structures or atop tall buildings, offer more distributed connection but without the upload-download speeds of WiFi (Barratt and Whitelaw 2011). To maintain their functionality, smartphones typically rely on both systems, always connected to cellular networks (assuming there is a signal) and automatically switching to WiFi when available, such as at home or work. This always-on pattern is required for a phone call or a text message to go through, but producing this ethereal connectivity effectively requires transforming the atmosphere into a send and receive broadcast medium powered by electricity. This is not new: radio has done the same for over 100 years (Mitchell 2003), but the sheer number of individual devices tethered into the system, the constant back and forth between user and network, and the standby nature of devices' constant broadcast all demand not-insignificant amounts of energy. Additionally, the mobile aspect of this ubiquitous connection necessitates each and every mobile device to be battery powered. In turn, those batteries must be charged regularly.

Transmitted through a WiFi router or cellular antenna site, the bits and bytes of information, for instance a request to load a website, typically travel through a local and then regional fiber-optic backbone. This backbone is most often routed under or alongside major transportation corridors, and especially in the United States, along railroad rights of way. The digitized information then arrives at a colocation node where a variety of mobile and terrestrial telecommunication providers transfer that information between their different networks and send it through the Internet backbone to the specific data center that holds the website (Wiig 2013). While the particular location of a data center does not matter to the user, only successful receipt of digitized information on their smartphone, there are specific, infrastructural, material places that hold the Internet, a point that will be returned to below. Returning the information to the user entails a reversal of the process. The energy demands of mobile devices and data are dynamic and anchored in expectations for constant connectivity (Bates et al. 2014; Spinney et al. 2012). In turn this demand necessitates regular charging of devices and ties into the above-mentioned cellular and WiFi networks, but also into the storage and transmission of data between a user's smartphone and the data center holding the information.

The connective, information-retrieval, social networking, and entertainment functionality of smartphones presents multimodal opportunities to spend time on these devices, and "patterns of use depends significantly on both location and social context" (Do et al. 2011: 2). Do et al. observe that devices like smartphones are used most at places other than home or work (where other computers were likely to be available), reflecting that "The use of [the mobile Internet] is highest at transportation-related places such as bus stop or train station. In such contexts, the Internet is likely to be used for reading news, looking for information or killing time" (Do et al. 2011: 5). Do et al.'s study purposefully did not track use of smartphones while in motion, nor did it measure when and where users charged their devices, but regardless the study identifies that using smartphones in transit complements use at home and at work, whether on a smartphone or another computing device like a laptop (Spinney et al. 2012). Other research on "human-battery interaction" builds off human-computer interaction research on digital interfaces, identifying two types of smartphone chargers: proactive users who habit-ually charge regardless of battery level and reactive users who charge depending on battery level (Rahmati and Zhong 2009). Users' expecta-tion was that a fully charged battery should last one day of smartphone use, leading to routines of charging overnight to have a full battery in the morning (Ferreira et al. 2015: 386). Most participants in Ferreira et al.'s study (2015) carried a charger cable and plug with them daily, and some augmented this by carrying an external battery pack. Considering charg-ing as a social practice, Horta et al. (2016) locate this mundane but nec-essary activity within the embedded habits and expectations that the smartphone *and* the wireless connective ICT network are always-on, as needed, and consequently, constantly demanding energy.

These above-mentioned studies indicate what can be qualitatively seen and experienced: that the majority of the population of a city like Boston or Philadelphia use smartphones a lot, especially while between places. While beyond the scope of this chapter, the manifold contexts of use as well as the practice of charging are particular and situated in larger, infra-structurally managed expectations of regular provision of connectivity and uninterrupted electricity supply, expectations that can go unmet at times in cities in the global South (Silver 2015). The more a smartphone is used, the more likely its battery will need to be charged before reaching home or work, leading to emergent practices around charging in public, either surreptitiously at a power outlet intended for another use, or at approved charging stations located in areas of high traffic such as train stations. These charging practices often extend onto long-distance trains and onto some commuter trains, where providing electricity for laptop computers has become an amenity to attract and maintain ridership (Amtrak 2013b).

Drawing inspiration from Watts' (2008) theorization of the time and space of train travel, this chapter unfolds as an ethnographic travel narrative drawn from fieldwork between 2013 and 2016, synthesizing numerous trips into one narrative journey. The narrative helps to reveal how energy demand is made. As described here, the co-produced dynamics of demand are assembled from multiple, layered relationships between obdurate and inanimate transport, telecommunication, and energy infrastructures, and emergent social practices around using and charging mobile devices while in transit, in motion, and in stations.

2.3   Expectations of Constant Connectivity and Public Charging: An Amtrak Rail Journey

After traveling from my home via a regional commuter train, I arrive at Boston's South Station for a trip to Philadelphia on the Amtrak Northeast Regional Railroad Corridor. This route is the busiest rail travel corridor in the United States, with 457 miles of track between Boston and Washington DC, carrying over 11 million passengers a year (Amtrak 2012, 2013a). As large cities in the northeastern United States, ones with diverse populations and economies, a journey between Boston and Philadelphia offers a quotidian case to contextualize the provision of charging and mobile connectivity. This case is presented to highlight the ubiquity and generalizability of rail travel and smartphone use within larger practices of mobility. Certain details would likely change in other locations in the global North, but the context offered in this chapter can be generalized across and between other cities and other railway journeys.

I am early for the train to Philadelphia, so I walk around the large, crowded station looking for a place to sit and check email on my smartphone. South Station is the main Amtrak station for Boston, the final stop for 8 of the region's 12 commuter trains, a stop for the subway and city buses, and adjacent to Boston's long-distance bus terminal. South Station's electrical outlets are covered and locked shut, preventing the public from accessing the plugs. While there are many tables to sit at, there is no *accessible* electricity provision. There are, however, three private companies providing charging services for a fee.

Around the world, ever-increasingly fast mobility relies on non-mobile, obdurate mobility systems (Urry 2007). A similar point can be observed here in the case of digital social exchange; with one important difference. Whereas mobility often relies on infrastructures with long lifespans, such as in the case of motorways, train stations, and airports, digital social exchange relies on infrastructures with both longer-term and shorter-term lifespans. For example, the oldest vendor, Gocharge, had a wall-mounted unit dangling a number of charging cords for popular phones right outside the men's bathroom that, when functioning, charges $3.00 for 25 minutes of charging (Fig. 2.1). As observed in fieldwork over the last three years, this unit is often broken and sporting a hand-drawn "OUT OF ORDER" sign. Additionally, the unit is in a high traffic area with no tables or seating nearby, so to charge their phone, a customer would have to stand in the way of men hurrying to and from the bathroom. As of fall 2016, the unit has been removed and no similar service has taken its place. In a more central area adjacent to the station's main food court sit two kiosks offering external smartphone batteries. The Mobile Qubes Portable Battery Rental offers, as the name implies, portable batteries for $4.99 to rent or $44.99 to buy. The kiosk has bright screens flashing advertising messages such as "NO OUTLET, NO CORD, NO PROBLEM". Directly adjacent to this kiosk is a smaller Morphie illuminated kiosk offering "universal batteries" similarly sized to a deck of cards or smaller, and protective cases with batteries built in for popular phone models like the Apple iPhone. Morphie's products cost between $35.00 and $129.00. All three kiosks are automated; the only way to pay for charging or to purchase a battery is with a credit card. South Station's largest concession is a two-level drug store that features a prominent display of smartphone cords and plugs for sale right before the checkout line, indicating the market for smartphone

Fig. 2.1 GoCharge's pay-per-use smartphone charging station in Boston's South Station. August 2013 (Photo by Alan Wiig)

charging equipment in transit and in general. The presence of these charging cables for smartphones as consumer goods, in addition to kiosks to rent or buy batteries and to buy charging time has replaced public telephones as a communication medium at train stations, with its attendant social practices and infrastructural presence within past patterns of mobility and connectivity.

Once on the Amtrak train, travelers can take advantage of the two electrical outlets at each pair of seats on Amtrak's passenger cars, free to use with the purchase of a ticket. As the train pulls out and the passengers get settled, seat-back trays release and laptops emerge from bags, in addition to smartphones and tablets and traditional paper-based media. These outlets were installed around 1998 (TrainWeb 1998). On longer journeys such as the six hours to Philadelphia, the outlets prove useful. Even if a device's battery will hold a charge to last the journey, people seem inclined to keep these objects topped off. One Yelp commenter praising Amtrak's provision of outlets writes "you have time to sit back and relax while charging your phone in the outlet next to your seat, score! The seats are huge so you have tons of personal space" (Yelp 2015), one indication of the service's usefulness to passengers.

The train passes through suburban Boston, running alongside the Atlantic Ocean in Connecticut, chugs through metropolitan New York, then enters the marshy wetlands of northern New Jersey before finally crossing the Delaware River and entering the post-industrial fringe of Philadelphia. The scene inside the train retains a similar feel even as passengers come and go: people pass the time on laptops, smartphones, reading books or magazines, eating, or looking out the window. Digital devices have, to an extent, replaced books and newspapers as a means of passing the time in transit. In a sense this is an evolution of the habit of reading, working, and generally passing the time in motion. Schivelbush in *The Railway Journey* mentions "the habit of reading while traveling was not only a result of the dissolution and panoramization of the outside landscape due to velocity, but also the result of the situation inside the train compartment" where "reading became a surrogate for the communication that no longer took place as it did in hired coaches before the advent of rail travel" (1986: 67). Reading on the Internet while on the train, however, involves maintaining a connection to the Internet, a process that has been fraught with complications.

As important if not more important than the ability to charge is Amtrak's provision of wireless Internet, although to continually use the Internet eventually requires charging a battery; the two operate together, underscoring the co-production of digitized social exchange. WiFi was added to the faster Acela business class trains in 2010 and the regional trains in 2011 (Berman 2011). An unobtrusive sticker on the window by each seat states: "Your seat is now a hotspot: WiFi on this train" (Fig. 2.2). However, the strength of the WiFi signal is dependent on the trackside cellular coverage that transmits the data from the moving train into the terrestrial digital infrastructure, coverage that varies due to location, environmental interruptions like a hillside or dense forest, the blanketing strength of the signal, and the number of users on the train. Streaming

Fig. 2.2 On an Amtrak Northeast Corridor train a sticker on the window states "Your seat is now a [WiFi] hot spot." November 2013 (Photo by Alan Wiig)

video websites like YouTube, which have high bandwidth needs, are blocked; Amtrak states: "Our WiFi bandwidth supports general web browsing, not streaming music and video, or downloading large files" (Amtrak 2014).

This level of connectivity was not reliable enough for some travelers nor the media outlets based, not coincidentally, on the Northeast Amtrak corridor, such as in New York City. While batteries can supplement laptops and smartphones if the electricity to the train car goes out briefly, there inherently can be no local backup of the Internet. Immediately upon initiating the service, complaints about spotty Internet coverage emerged in the media. *The New York Times* titled an article "Wi-Fi and Amtrak: Missed Connections" (Nixon 2012) and *The Economist* complained that the "Wi-Fi should actually work" (The Economist 2011). The expectation implicit in these two articles is that travelers, especially business travelers or politicians passing between New York City and

Washington DC, must have constant connectivity or they will find other means of travel. Commentary such as this fosters and reinforces expectations of WiFi and charging that in turn contributes to the social, technological, and infrastructural co-production of mobile connectivity. Consequently, in 2014 Amtrak sought a contractor to build a trackside wireless network alongside the Northeast Corridor by 2019 (Koebler 2014; Meyer 2014). The new network will be comprised of base stations near the rail line that will connect trains via fiber optic lines or radio-based connectivity, building a wireless network just for its trains and passengers instead of sharing existing cell towers that weren't built to support speeding trains filled with travelers demanding connection. In the meantime, it seems more passengers attempting to access the network has reduced connectivity even more (Tangel 2015). For many travelers, train travel on this Amtrak corridor cannot be separated from the presumption of constant digital connectivity as well as access to electricity, and the absence of a perfect, high-speed connectivity leads to significant complaints. Considerations of charging smartphones and laptops, as the end-user point of energy demand, are located within larger electrical infrastructure assemblages powering telecommunication systems: the Internet and mobile communication.

Once at Philadelphia's 30th Street Station, I have a few minutes to wander around and stretch my legs before my local commuter train to Temple University arrives. Regarding charging, Boston's South Station as mentioned above has privatized, fee-based charging or none at all, while at 30th Street Station, Amtrak provides charging services to travelers for free. A "Cell Phone Charging Station" with numerous outlets along a wooden bench is tucked alongside the Amtrak ticket counter. The electricity outlets in the public areas of the terminal are open, even though they are at floor level in a pedestrian corridor, leading travelers in need of electricity to crouch or hunch with their baggage, transforming their posture in order to charge and use their smartphones (Fig. 2.3).

My regional train arrives so I board for the short, 15-minute trip across Philadelphia's downtown. The commuter railcars do not provide outlets for charging, though there are outlets for the cleaning and maintenance of the trains. Here I see my final example of charging practices, with a Temple University student sitting in the corridor-

Fig. 2.3 A waiting traveler hunches over his smartphone plugged in to an electricity outlet in Philadelphia's 30th Street Station. April 2015 (Photo by Alan Wiig)

side seat by the door in order to access the outlet and charge his smartphone before, I assume, classes. He is resting, eyes closed and head down on crossed arms, but phone sitting on his lap, plugged in. Even on a relatively short journey, and before spending time on a university campus with many opportunities to charge a battery, the student was still compelled to use this informal source of charging (Fig. 2.4).

2.4   Data Centers and the Electricity Underlying Digital Connectivity

On the commuter rail trip through Philadelphia I pass by one of the largest data centers in the Northeast United States; this physical proximity offers a narrative tie-in to finish this chapter by returning to the most

Fig. 2.4 A passenger uses an outlet on a Philadelphia commuter rail train to charge his smartphone while he rests. November 2013 (Photo by Alan Wiig)

prominent source of energy demand of the Internet: the data centers that hold the information. This energy demand is not insignificant:

> Direct electricity used by information technology equipment in data centers represented about 0.5% of total world electricity consumption in 2005. When electricity for cooling and power distribution is included, that figure is about 1%. Worldwide data center power demand in 2005 was equivalent (in capacity terms) to about seventeen 1000 MW power plants. (Koomey 2008: 1)

Furthermore, "Worldwide electricity consumption for all communication networks [is] 1.8% of total energy use" and growing 10% per year (Lambert et al. 2012: 1). These points are underscored by the recognition that: "As of 2007, the average data center consumed as much energy as

25,000 homes. There are at least 5.75 million new servers deployed into new and existing data centers every year. Data centers account for at least 1.5% of US energy consumption and demand is growing 10% per year" (Bartels 2011). Globally, these information warehouses use about 30 billion watts of electricity, roughly equivalent to the output of 30 nuclear power plants (Glanz 2012). Interestingly and importantly when considering demand for energy, data centers "were using only six percent to twelve percent of their electricity powering their servers to perform computations. The rest was essentially used to keep servers idling and ready in case of a surge in activity that could slow or crash their operations" (Glanz 2012). *Maintaining information at standby, because of the expectation of near-instant access, accounts for essentially 90% of a data center's energy demand.* Ensuring accessible, digital social exchange is inseparable from regional electrical power grids. Such obdurate, non-mobile grids are a very grounded, equipment-heavy, and resource-intensive infrastructure. This infrastructure is not just data centers, but includes network equipment, fiber-optic cabling, and cellular antenna sites, and the electrical energy infrastructure that powers these interconnected systems.

Underscoring the obduracy of mobile connectivity, the social practices discussed in this chapter rely on constant connectivity to data centers. The location of data centers does not inherently matter, only the connection. Data, through its embeddedness in infrastructure, embodies material spaces, forming an infrastructural geography layered within and between cities. This infrastructural geography is co-produced through and alongside transportation networks, especially the railroad, even within city streets where the railroad has been absent for decades, and functions through massive amounts of electricity. The Terminal Commerce Building that I passed on the way to Temple University is the largest data center in Philadelphia: an 11 story, 1.3 million square feet structure located just north of downtown Philadelphia, in view of City Hall. It is an Art Deco building that takes up an entire city block, completed in 1930 and originally containing wholesale furniture showrooms and warehousing for furniture. The structure has a railroad spur going into the basement to load and unload goods; this is where the fiber optic cables run into the data center; this otherwise unused railroad right of way cutting through downtown is what makes the building useful as a data center. According to a report submitted to register the building on the National Register of Historic Places, a status granted in 1996, the Terminal Commerce Building was "large enough to command its own post office, and later its own zip code, [and] by 1948 the building housed 175 companies employing some 5,000 people" (National Park Service 1996: 6). By the early 1970s this was the largest single office building in the city. By the late 1990s the building was modified into a co-located data center where multiple private carriers house their respective data centers in one building. Today there are approximately 80 carriers in the building, which also acts as an interchange between data storage providers and telecommunications companies (DiStefano 2014; Verge 2014). The more than 5000 people who originally worked in the building have now been replaced with thousands of inanimate data servers. The energy demanded by the data center manifests in an electricity sub-station located across two blocks behind the data center. The presence of the sub-station highlights how much energy is consumed to power both the building and the digital services within, so much that it links directly into the power grid itself.

The energy demand of data centers, as well as their obdurate embeddedness in the urban landscape, form the infrastructural back end in the co-production of mobile connectivity. The mobility of train travel relies on the fixity of tracks, stations, and energy sources to power the trains. Adding a layer of wireless digital connectivity to complement train travel necessitates not only the provision of connectivity and charging on the train, but also entangles a whole digital infrastructure with its own energy demands and material geography.

2.5 Conclusion

This chapter has drawn from an ethnographic study of digital social exchange to reveal the dynamics of energy demand. Increasingly mobile, digital, social exchange is bound up in infrastructures, expectations of constant connection, and emergent charging practices. Demand for energy requires different sites and material arrangements, and support from new expectations and conventions around location and availability for energy. Applying Jasanoff's theory of co-production creates a framework for conceptualizing mobile connectivity through the multiple infrastructures that underlie this social practice. Co-production in this chapter's case involves emergent social practice of mobile connectivity in motion enabled through smartphones and the telecommunication infrastructure wirelessly linking devices to the Internet but also reliant on the regular charging of the devices batteries, which in turn are producing related practices of charging in public. Charging has, as discussed in Boston's South Station, also opened up business opportunities to provide travelers with mobile batteries in order to maintain their device's power and, consequently, its connective capacity.

As an emergent social practice, mobile connectivity cannot function without electricity; it has a distinct energy demand stretching from, in the case provided in this chapter, the collectives of individual travelers using the mobile Internet and charging their smartphones in motion, to the data centers and cellular infrastructure propagating the digitized connectivity through the ether. With regard to the creation of energy demand, the co-production of mobile connectivity is entangled within short- and long-term transformations within infrastructure, with regard to the alteration of existing buildings into data centers, institutions such as Amtrak which had to evolve into a telecommunications provider in addition to its established role as a railroad transportation provider, and businesses such as Morphie or Gocharge, both relatively new companies providing energy-related services that maintain social practices around mobile connectivity. The expectation of connectivity on Amtrak's Northeast Corridor also includes supplying electricity to power mobile devices throughout one's journey. As an emerging social practice, charging mobile devices in public is built around the assumption and expectation that energy for smartphone batteries is either a traveler's right, such as in Philadelphia, a business opportunity, such as in Boston, or a factor of the rail provider retaining passengers who might otherwise drive or fly.

The energy required to power mobile devices, network equipment, and data centers is inseparable from mobile, digital social exchange. Demand for standby, "always-on" wirelessness (Mackenzie 2010) entails a constant provision of wireless signals pulsing between smartphones and base stations, transnational telecommunications systems, and data centers scattered around the world. These inanimate, obdurate systems of connectivity have a particular, material arrangement stretching beyond each user that is produced through and grounded in electricity.

Transformation of energy demand has occurred alongside shorter-term transformations in the routines and patterns of everyday lives. While a traveler's focus is on maintaining their linkage to the Internet, the underlying demand for electricity to power mobile devices speaks to a continuous and growing appetite for energy fundamental to the digitization of society in the twenty-first century. This digitization and its energy demand are transforming space: the dynamics of connectivity alter access to energy, especially in public spaces where such access in the past was not necessary nor expected. This all speaks to the complexities involved in managing transportation systems today, complexities that extend into and are created by everyday social practices, in particular, practices connected to digitized social exchange.

Acknowledgements The research underlying this essay was supported by an international research fellowship at the DEMAND Centre at Lancaster University in 2015 as well as a post-doctoral research fellowship in Temple University's Office of the Vice President for Research, under Dr. Michele Masucci. Thanks to the scholarly community at the DEMAND Center and in particular Elizabeth Shove, Gordon Walker, and Allison Hui for conversations, encouragement, and advice. Lastly, thanks to the reviewers and editors for their comments, critique, and feedback.

Notes

1. The release of the Apple iPhone in 2007 was widely seen to herald the widespread adoption of smartphones and the concurrent need for telecommunications and mobile communication providers to offer the high-speed wireless, mobile Internet service smartphones rely on (Arthur 2012). The digital infrastructures facilitating mobile connectivity, and the social practices resulting from this connectivity—as well as the energy demand stemming from the practices—are consequently all very recent, less than a decade old.

Bibliography

Amtrak. 2012. *The Amtrak vision for the Northeast Corridor: 2012 update report*: Amtrak. Available at: https://www.amtrak.com/ccurl/453/325/Amtrak-Vision-for-the-Northeast-Corridor.pdf

———. 2013a. Amtrak sets ridership record and moves the nation's economy forward: America's Railroad helps communities grow and prosper. Amtrak. Available at: https://www.amtrak.com/ccurl/730/658/FY13-Record-Ridership-ATK-13-122.pdf

———. 2013b. *Go girl! And take these 5 accessories with you*. Available at: http://blog.amtrak.com/2013/08/travel-accessories/

———. 2014. *Tweet 9th June 2014, 11:13 a.m.* Available at: https://twitter.com/Amtrak/status/476064573171052544?ref_src=twsrc%5Etfw

Anderson, M. 2015. *Technology device ownership: 2015*. Available at: http://www.pewinternet.org/2015/10/29/technology-device-ownership-2015/

Arthur, C. 2012 The history of smartphones: Timeline. *The Guardian*, January 24. sec. Technology. Source: https://www.theguardian.com/technology/2012/jan/24/smartphones-timeline. Last accessed 11 Nov 2016.

Barratt, C., and I. Whitelaw. 2011. *The spotter's guide to urban engineering: Infrastructure and technology in the modern landscape*. Buffalo: Firefly Books.

Bartels, A. 2011. [INFOGRAPHIC] Data center evolution: 1960 to 2000. *Rackspace blog*. Available at: https://blog.rackspace.com/datacenter-evolution-1960-to-2000. Accessed 9 Mar 2017.

Bates, O., M. Hazas, A. Friday, et al. 2014. Towards an holistic view of the energy and environmental impacts of domestic media and IT. *Proceedings of the 32nd annual ACM conference on Human factors in computing systems.* ACM, 1173–1182.

Berman, M. 2011. Amtrak adds free Wi-Fi to a dozen routes. *The Washington Post* – Blogs. http://www.washingtonpost.com/blogs/dr-gridlock/post/amtrak-adds-free-wi-fi-to-a-dozen-routes/2011/10/31/gIQAdXeKZM_blog.html. Last accessed 18 Feb 2015.

DiStefano, J. 2014. $70M upgrade to telecom hotel at 401 N. Broad. *Philly.com*. Accessed 9 Mar 2017.

Do, T.M.T., J. Blom, and D. Gatica-Perez. 2011. Smartphone usage in the wild: A large-scale analysis of applications and context. *Proceedings of the 13th international conference on multimodal interfaces*, 353–360. ACM.

Ferreira, P., M. McGregor, and A. Lampinen. 2015. Caring for batteries: Maintaining infrastructures and mobile social contexts. *Mobile HCI '15*, Copenhagen, 383–392.

Glanz, J. 2012. Power, pollution and the internet. *The New York Times.* Available at: http://www.nytimes.com/2012/09/23/technology/data-centers-waste-vast-amounts-of-energy-belying-industry-image.html. Accessed 9 Mar 2012.

Horta, A., S. Fonseca, M. Truninger, et al. 2016. Mobile phones, batteries and power consumption: An analysis of social practices in Portugal. *Energy Research & Social Science* 13: 15–23.

ITU. 2014. *Mobile-broadband uptake continues to grow at double-digit rates.* International Telecommunications Union (ITU). Available at: https://www.itu.int/en/ITU-D/Statistics/Documents/facts/ICTFactsFigures2014-e.pdf

Jasanoff, S. 2004. *States of knowledge: The co-production of science and social order.* London: Routledge.

———. 2005. *Designs on nature: Science and democracy in Europe and the United States.* Princeton: Princeton University Press.

Koebler, J. 2014. Amtrak is becoming its own ISP. *Motherboard.* Available at: https://motherboard.vice.com/en_us/article/amtrak-is-becoming-its-own-isp. Accessed 9 Mar 2017.

Koomey, J.G. 2008. Worldwide electricity used in data centers. *Environmental Research Letters* 3 (3): 1–8.

Lambert, S., W. Van Heddeghem, W. Vereecken, et al. 2012. Worldwide electricity consumption of communication networks. *Optics Express* 20: B513–B524.

Mackenzie, A. 2010. *Wirelessness: Radical empiricism in network cultures.* Cambridge, MA: MIT Press.

Meyer, R. 2014. You might soon be able to watch Netflix on Amtrak Wifi. *The Atlantic.* Available at: https://www.theatlantic.com/technology/archive/2014/06/you-might-soon-be-able-to-watch-netflix-on-amtrak/372435/. Accessed 9 Mar 2017.

Mitchell, W.J. 2003. *Me++: The cyborg self and the networked city.* Cambridge, MA: MIT Press.

National Park Service. 1996. *National register of historic places registration form: Terminal commerce building.* Washington, DC: United States Department of the Interior. Source: https://www.dot7.state.pa.us/ce_imagery/phmc_scans/H097510_01H.pdf. Last accessed 25 Mar 2015.

Nixon, R. 2012. Wi-Fi and Amtrak: Missed connections. *The New York Times.* Available at: http://www.nytimes.com/2012/06/03/travel/wi-fi-and-amtrak-missed-connections.html. Accessed 9 Mar 2017.

Rahmati, A., and L. Zhong. 2009. Human–battery interaction on mobile phones. *Pervasive and Mobile Computing* 5: 465–477.

Schatzki, T. 2010. Materiality and social life. *Nature and Culture* 5: 123–149.

Schivelbusch, W. 1986. *The railway journey: The industrialization of time and space in the 19th century.* Berkeley: University of California Press.

Shove, E., G. Walker, D. Tyfield, et al. 2014. What is energy for? Social practice and energy demand. *Theory, Culture & Society* 31: 41–58.

Shove, E., M. Watson, N. Spurling, et al. 2015. Conceptualizing connections: Energy demand, infrastructures and social practices. *European Journal of Social Theory* 18: 274–287.

Silver, J. 2015. Disrupted infrastructures: An urban political ecology of interrupted electricity in Accra. *International Journal of Urban and Regional Research* 39: 984–1003.

Spinney, J., N. Green, K. Burningham, et al. 2012. Are we sitting comfortably? Domestic imaginaries, laptop practices, and energy use. *Environment and Planning A* 44: 2629–2645.

Tangel, A. 2015. Wi-Fi Woes Plague Amtrak on Northeast Corridor. *The Wall Street Journal*. Available at: https://www.wsj.com/articles/wi-fi-woes-plague-amtrak-on-northeast-corridor-1447033522. Accessed 9 Mar 2017.

The Economist. 2011. *Wi-Fi should actually work*. Available at: http://www.economist.com/blogs/gulliver/2011/12/wireless-internet-amtrak

TrainWeb. 1998. *Electric outlets on Amtrak*. Available at: http://www.trainweb.com/travel/general/electric.htm#sthash.sn3GPyN8.dpbs

Urry, J. 2007. *Mobilities*. Malden: Polity.

Verge, J. 2014. New owners for Philadelphia carrier hotel at 401 North Broad. *Data Centre Knowledge*. Available at: http://www.datacenterknowledge.com/archives/2014/03/06/new-owners-philadelphia-carrier-hotel-401-north-broad/. Accessed 9 Mar 2017.

Watts, L. 2008. The art and craft of train travel. *Social & Cultural Geography* 9: 711–726.

Wiig, A. 2013. Everyday landmarks of networked urbanism: Cellular antenna sites and the infrastructure of mobile communication in Philadelphia. *Journal of Urban Technology* 20: 21–37.

Yelp. 2015. *Comment by Colleen B, 22nd February 2015 – 30th street station*. Available at: https://www.yelp.com/biz/30th-street-station-philadelphia-2?hrid=f8OjSH0UU9TCP0-ol7rKAQ&utm_campaign=www_review_share_popup&utm_medium=copy_link&utm_source=(direct)

Alan Wiig is an Assistant Professor of Urban Planning and Community Development at the University of Massachusetts, Boston, USA. His research has examined the infrastructure of Internet and mobile communication, as well as smart urbanization: the transformation of urban governance, civic exchange, and economic development through information technology. His current research project analyses the geopolitical underpinnings of global infrastructures underlying urban regeneration across the North Atlantic.

3

Constructing Normality Through Material and Social Lock-In: The Dynamics of Energy Consumption Among Geneva's More Affluent Households

Marlyne Sahakian

3.1 Introduction

Energy consumption is tied to almost all aspects of everyday life, from cooking to cleaning to sleeping comfortably indoors. Yet how and why people use energy services can vary greatly across cultures and socio-economic groups. These variations have to do with how social practices play out, involving different dimensions of a practice—for example, people's skills and competencies, the material dimension of consumption and socio-cultural contexts (Sahakian and Wilhite 2014). Ground-breaking work on social practices has focused on how the social norms of comfort, cleanliness and convenience (Shove 2003) hold together resource-intensive practices. Social norms, in the Durkheim sense, can be understood as moral rules which, if not followed, could result in social sanction and reprobation. Who delivers this social sanction, and whether sanctions differ in varying socio-cultural contexts, remains an empirical

M. Sahakian (✉)
University of Geneva, Geneva, Switzerland

© The Author(s) 2018
A. Hui et al. (eds.), *Demanding Energy*, DOI 10.1007/978-3-319-61991-0_3

question. Picking up from Bourdieu, people's everyday practices are shaped by different forms of capital, acquired socio-historically, in addition to people's dispositions and the social fields within which they act (Bourdieu 1979: 331). If people's dispositions include feelings, beliefs, competencies and expectations around everyday life, these dispositions also relate to the notion of normality—how people come to expect, feel or understand what is "normal", which could also differ across varying contexts.

A central theme in this chapter is how expectations around social norms, or *normality*, are constructed, and the role of social reprobation in shaping these expectations. To explore this theme, we focus on the electricity consumption of affluent households in Geneva, Switzerland, who self-identify as being part of the expatriate or "expat" population.[1] The goal of focusing on this group is not to put forward a moral argument, that solely so-called elites should be targeted in relation to current consumption patterns: all sectors of society and all socio-economic groups have a role to play towards absolute reductions, in the context of industrialized countries. Rather, we consider these consumers as a worthwhile area of study because of the financial, social and cultural capital at their disposal. Mobility is one signal of these forms of capital, adding another dimension to the analysis: how normality and social reprobation differ in varying spaces of consumption around the world. In the first phase of the "practice turn" in consumption studies (in the early 2000s), there was little consideration for consumption across different socio-economic groups, which was an integral part of Pierre Bourdieu's earlier interpretation of practices (see Dubuisson-Quellier and Plessz 2013). This research gap is being filled with a growing body of work considering consumption practices in relation to social class and power (see Anantharaman 2016; Plessz et al. 2016), and this chapter seeks to contribute further insights.

Regarding how consumption practices could change over time towards less resource intensive practices, one hypothesis is that this would require the breaking down of more than one element of a practice (Sahakian and Wilhite 2014). It would follow that maintaining certain practices in place would involve either avoiding any weakening of the elements, or conversely, a strengthening of certain elements that hold a particular practice together. With the financial means to live in large and multiple homes and acquire a large range of appliances, the case of affluent Geneva house-holds presents an interesting study in relation to the material dimension of practices, but also with respect to belonging to a social group, with high social and cultural capital. To understand the significance of these aspects in maintaining certain practices over time, we engage with the notion of "lock-in"—traditionally used to define the path dependency of certain technologies which, even when inferior to alternative technologies, are maintained because they follow paths that are difficult or costly to escape from. Rather than consider technological lock-in, this chapter proposes *material and social lock-in* as broader concepts, in order to further explore the role of things and socio-cultural contexts in constructing normality over time.

What follows is an overview of main concepts guiding the research, as well as the methodology. The research findings are presented in relation to three main themes: material lock-in, or how appliances, things, infrastructures and spaces can promote a certain interpretation of practices over others; social lock-in, involving the setting of expectations in relation to reprobation or encouragement from social groups; and finally, the locking-in or, in some cases, the un-locking of certain practices across different contexts and cultures, taking into account the mobility of this group. In the conclusion, opportunities for destabilizing energy-intensive practices are discussed.

3.2   Concepts and Methodology

While social practice theories have been elaborated as of late in relation to (un)sustainable consumption (Shove 2003; Warde 2004; Sahakian and Wilhite 2014), Bourdieu's (1979) earlier work on food consumption practices in France was more focused on questions of power and class relations. Based on ethnographic methods, Bourdieu proposes the notion of total capital resources, to include not only economic capital, but also cultural capital (related to schooling), social capital (related to social networks) and symbolic capital (related to reputation and competencies). People's everyday practices are shaped by these forms of capital, acquired through life experiences, in addition to people's dispositions and the social fields within which they act (Bourdieu 1979: 331). Bourdieu organizes social life into these social fields—such as political, economic, or cultural fields, with participants in a field sharing the same rules of the game, or a tacit understanding of the stakes at hand, which translates into a practical understanding of the rules that govern a social field (Bourdieu and Wacquant 1992: 93). Not all participants have the same position within that field, leading to social pressures or relations of power. The domestic sphere can also be apprehended as a social field, involving people in a similar social class who are aware of the standards governing how to act, what to say and even what to feel in relation to home-making, involving forms of encouragement and reprobation.

Similar to Bourdieu's notion of disposition is Schatzki's understanding of teleoaffective structures, which includes ends, projects, tasks, purposes, beliefs, emotions and moods (Schatzki 1996, 2002). As part of an enactment of a practice, the role of emotions in relation to teleoaffective structures merits further exploration in sustainable consumption studies (Spaargaren 2011) and is currently understudied in the literature. In considering expectations, we engage with the sociology of emotions methodologically (Flam and Kleres 2015) to uncover how emotions emerge in practices, involving doings and sayings, and how this relates to normality—or what people come to expect as normal. This research builds on earlier work by the author, exploring the role of guilt and shame in recruiting new practitioners to practices that involved household appliances, in post-World War Europe (Sahakian 2015). Interviews were coded in relation to how the interviewer interpreted emotive responses, including feelings of entitlement, embarrassment or indifference. This emotional reading of the transcripts and interview notes allowed us to further explore the question of norms, specifically, how people feel that they are either adhering to a norm or infringing upon a social norm—involving socially acceptable or reproachable practices. Feelings of entitlement have been studied around air conditioning in Singapore, and how certain groups come to expect artificially cooled air as a norm (Hitchings and Lee 2008). We also accounted for feeling rules (Hochschild 1979), or the recognition that how people express themselves emotively is also guided by social norms, which can lead to controlled emotive responses.

For this chapter, practices are seen as being held together by three elements—people and their dispositions, including teleoaffective structures, the socio-cultural context, involving social norms, and the material dimension of consumption, including technologies, but also artefacts and infrastructure (building on Sahakian and Wilhite 2014). In relation to these elements, this chapter introduces the notion of social lock-in, or how social groups enforce the upkeep of norms, as well as material lock-in, involving the material dimension of practices. Since the late 1980s, technological lock-in has been used to explain how technological paths of development can be costly or difficult to change due to historical or economic factors, leading to path dependence for inferior technologies (Arthur 1989). In the 1990s, technological lock-in was placed in relation to environmental change, towards understanding the consequences of technological path dependency in relation to resource usage (Rip and Kemp 1998). A more recent study in this vein considers how the institutional conditionings guiding economic markets can lock in one economic paradigm over another, leading to institutional lock-in (van Griethuysen 2010). These approaches tend to focus on the macro-scale, highlighting "how ecological change is deeply embedded in complex, interdependent technological and socio-economic systems" (Perkins 2003: 6). The term "moral lock-in" has also been applied to studying path dependency in relation to technologies that promote morally inferior pathways, linked to ethics and social responsibilities (Bruijnis et al. 2015). For this chapter, the scale of analysis is shifted to the household: demonstrating how people can be "locked into" certain consumption practices by their physical possessions, which can include technological appliances but also the number and size of their homes. The term "material lock-in" is therefore used as more encompassing than technological lock-in. The role of social norms in holding together certain practices, carried by people in social groups, is positioned as a form of "social lock-in", related to social status and power dynamics within set social fields involving family and peer groups. While recognizing that practice theory and innovation studies can be ontologically distinct, the concepts of material and social lock-in do not necessarily relate to paths of development in this case study, but rather to further understanding the interactions between people, social norms and the material dimension of consumption in the ongoing construction of normality.

This chapter is based on data drawn from a broader research project, studying electricity-related consumption practices among differing socio-economic groups in Lausanne and Geneva, Switzerland. Ten in-depth interviews were conducted in households, among women who self-identify as being part of the expatriate population in Geneva. Two of the women work full time outside of the home, two women do so part time and a fifth woman was working in the financial sector and was job searching at the time of the interview. The other five women do not have professional jobs, but work in the domestic sphere caring for children. All of the women have children, ranging from babies to teenagers. The women come from different backgrounds: South Africa, India, the United Kingdom, Jordan and several from the United States, but all have lived in several countries before arriving in Switzerland. We used vignettes and scenarios in interview situations, or stories about hypothetical people in a given situation, which respondents were asked to react to. This method allowed for some distancing between the respondent and the social norms we were seeking to discuss in relation to cleanliness, connectivity and gendered relations in the home, among others. The interviews were transcribed and, along with observations and notes, coded using NVivo 10 software. Emotional responses were captured in our observations and also directly in the transcripts of the interviews through coding (for example, laughter, hesitation, strong statement, strong positive feeling and strong negative feeling). In the transcript extracts used below such additions are noted in square brackets. Recruitment was through a snowballing technique. First contacts were made through an acquaintance who organized a social event where we were able to introduce this research project and recruit participants. Four women agreed to be interviewed in the first phase, which then led to six additional interviews. All of the interviews, save for one, were conducted in English. Quotes from the French interview, which often included English statements, have been translated. All of the names have been changed, to protect anonymity.

3.3   Research Findings

In order to understand how Geneva expats come to have expectations around what is a normal level of consumption, the sections below present empirical research on how normality is put into practice in three ways: first, how norms develop in relation to materiality; second, the role of social groups in setting expectations; and third, how norms are adapted and negotiated in different contexts.

Material Lock-In: Appliances and Spaces

The Geneva expats we met with live in large homes or spacious apartments. Some have swimming pools, heated for the most part, as well as Jacuzzis, saunas, garage space for at least two cars, multiple bathrooms and bedrooms, and large kitchens. Interior design is important, as homes are carefully outfitted in styles based on taste preferences and current trends—for example, towards more open-style kitchens that communicate more directly with living and dining areas. Electrical appliances and information communication technologies (ICTs) of all sorts are apparent throughout the homes, from kitchens outfitted with American-style refrigerators and flat-screen televisions, to the film projection rooms and entertainment areas, and multiple tablet, mobile and desktop computers.

Refrigerators are generally large in size, comparable to what can be found in North American homes; the exception was in homes that were rented and not owned,[2] where refrigerators more adapted to Swiss standards had been installed by the property owners. In this case, a second or third refrigerator is considered a necessary investment. Both Sally and Helene are in this situation; their husbands work in the financial sector in Geneva, and they are renting their homes. Sally explained that they bought a second refrigerator and freezer, which they keep in the basement, because:

> Well my husband, we have a very small fridge here (…) You know it's a half freezer it's a half fridge and it's European style. It's incredibly small, I'll show you [footsteps]. So this is a classic non-American fridge [laughter]. Well you can see it's like, it's minuscule.

The 180 cm-high refrigerator is a one-door unit, quite standard and even large for certain Swiss homes, with 90 cm-high refrigerators. Experiences from elsewhere, in this case the United States, make the European model now appear small in comparison—as further discussed in the third section below.

Helene purchased a new refrigerator as a result of an American Thanksgiving dinner, when she did not have sufficient space in her refrigerator and was storing food outside:

> I mean, it was so embarrassing, like we would be, I mean, putting things outside and it was [giggles] it wasn't cold enough, it was difficult to store things. If company would come and like stay for longer than a week at a time, it's easier if you cook in advance, but you need to have a place to store it.

The convenient access to cooled items was also mentioned in relation to a new feature: the refrigerated, under-the-counter drawer system, which one woman had installed in her kitchen, in addition to the existing refrigerator in the kitchen, and spare refrigerator in the basement. These under-the-counter models can be more energy efficient, as they often include at least two units, or two refrigerated spaces which can be kept at different temperatures adapted to the food or beverages being stored; they also release less cool air, with smaller openings than a large-sized refrigerator. In a review of promotional materials around these new appliances, the aesthetic appeal is put forward: the drawers are said to seamlessly integrate into kitchen design and are hidden from view, in opposition to the mass of a more standard refrigerator and freezer unit. In this case, however, the drawers are not replacing an existing refrigerator, but are an added feature—we can therefore note the multiplying of food and beverage cooling devices, in certain households.[3]

Another kitchen appliance that is multiplying is the dishwasher. Mariana tells me about her two dishwashers in the kitchen (see Fig. 3.1), as something that she knows could be considered excessive by some, but that she also cannot live without, as she put it: "I don't know how I managed before. I just can't imagine only having one dishwasher now". She has a similar installation in her chalet kitchen, in the ski resort of Verbier.

Fig. 3.1 The doubling of dishwashers in an affluent home

For this family, the double dish-washer has become legitimate, related to norms around tidiness and cleanliness: dirty glassware, dishes and pots no longer have to wait their turn on the counter, they can all be cleaned up simultaneously, particularly after social events. The dishwashers become a form of demonstration, which influence the degree of acceptability of these double appliances by people visiting this home, from among their social network.

New refrigerators, freezers and dishwashers are purchased during a redesign of the kitchen and because of financial means, but also because there is space for them. In turn, limits to appliance acquisition are directly related to available space—a form of material lock-in which can both limit and promote technological acquisition and multiplication. Helene, for example, would have liked to acquire more appliances but feels her counter space is too small. Having more, rather than less, is generally seen as desirable when it comes to living spaces. "Space, space to me is the ultimate luxury [Strong statement] (...). This feels too small to me now," explained Mariana, mother of two, talking about her three-storey attached house, with three bedrooms, living room and dining area, an office and basement area. Her main desire was to have additional bedrooms, particularly for welcoming guests.

Appliance acquisition and usage are separate matters, when it comes to electricity consumption. Having sufficient space to acquire new appliances does not mean that they are actually used or consuming electricity. When asked what appliances she has in her home, Mariana responds:

Interviewee: We have every appliance here. And I mean, seriously, every one.
Interviewer: Like what, do you have one of those knives that cut electronically?
Interviewee: Probably [laughter], no I'm serious.
Interviewer: And do you use them all very often?
Interviewee: No I have stuff I don't even... We have a... We have everything... I have a popcorn machine, I have a vacuum pack machine, I have a machine to cook things on bain marie [water bath for slow cooking], probably two...
Interviewer: What's the vacuum pack machine for?
Interviewee: You know when you want, when you want to seal something and you want to vacuum pack it and then you want to cook the meat in water that's at 90 degrees, I have all that. I have a blender, I have a hand blender, I have a Nutri Bullet, I probably have two waffle irons.

Hindering usage is the amount of the time needed for cleaning and maintaining the appliances. As Sally, mother of two, put it, "using appliances is a waste of time, then you have to wash them". Saskia, who has a three-year-old daughter, explains that she would much rather do things by hand, such as pressing lemons or grating carrots, than bother to take out electronic devices, use, then clean and store them.

One area that is dematerializing is that of entertainment, as ICTs make it possible to share and store books, music and films, in a manner that is more convenient for some as it implies less "hard copy" books, CDs and DVDs to store in the home. Time and space are both saved, through digital technologies. Sally explains that her uncle's collection was not made useful after his death and that books would have to be "extremely rare", as she put it, to have any true value. She has transferred her book collection to her iPad.

The people interviewed for this study know that we are researching energy consumption, which presents a bias: they are willing to explore their energy consumption in the domestic sphere with someone in the academic sphere. In that respect, it is no surprise that specific features of the home are presented with more of an apologetic tone, indicating a feeling of excess. For example, Helene, mother of three, showed me the elevator in her four-storey home in an apologetic manner: "So this I'm sure, it's a little bit ridiculous [laughs]." She explains that she uses the elevator for bringing up groceries, from the garage to the kitchen, or for cleaning and tidying the home; toys for example will be placed in the elevator, to go from the common living space back up to children's bedrooms. The fact that an elevator was installed in this apartment means that it will be used, which represents a form of material lock-in to energy-intensive vertical movement.

The doubling of appliances and added features, such as the double dishwasher or the elevator, create forms of demonstration and display, which can lead to legitimization processes in relation to tidying and cleaning practices. This in turn leads to the setting of new expectations, particularly in relation to acceptance by social groups or social pressures, both forms of social lock-in, a theme we will now turn to.

Social Lock-In: Social Acceptance and Pressures

For all respondents, maintaining cleanliness and order in the home is a standard that is non-negotiable. Norms around cleanliness and tidiness are very much upheld, with routines established around regular laundry and cleaning activities. This group has high standards and the means to put them into practice, involving the execution of chores with the support of domestic helpers, including caretakers for the children, and cleaners responsible for a tidy home, and clean, ironed laundry.

Mariana laughs as she tells me that someone is constantly cleaning her house; when I ask why she's laughing, she explains. "Because she comes an obscene amount of hours because I like things very clean. And I, I think I probably have every day someone at least spend five hours a day cleaning the house." For Mariana, excessive cleaning is desirable—normality in relation to cleanliness seems to have no upper limits, in most of the households visited. Selma, mother of two small children, explains that the laundry runs almost continuously in her home. Melanie, mother of an eight-year old boy, explains that her domestic helper cleans clothes continuously, any clothes that are left around the house will be cleaned, rather than put away. This may have less to do with dirt than maintaining a position within the households, or being locked into employing domestic help, as she explains:

> I also have, I suspect – I don't know if this is true and it may be unfair but I think sometimes she sees it as part of her job security, that she – because we don't really need her as much as we have her but we pay for all those hours because we want… we don't want to lose her, so sometimes I feel like she makes a bigger deal out of the laundry because it justifies us having her for the hours we do.

All of the women interviewed in this study have a clear understanding of their social position: they recognize, in discourse, their privileged situation. The type of language they used tended to be assertive, with little hesitation in expressing strong opinions. This made it difficult, at times, to get beyond controlled emotive responses—what has been termed feeling rules (Hochschild 1979). Being a privileged woman managing a large home also means that there is a sense of entitlement around the type of domestic help you can expect, but also the cleanliness, tidiness and indeed cosiness that you must then deliver for other household members. As Melanie explains:

> Thankfully we're, you know – my husband is providing a lifestyle in which we have help because I would not be happy if I did not have Dorine [the cleaning lady] running this big a house. When we were buying a house, I said to him, this is too big for us, you know. We don't need all this space, and he was, like, it'll be fine, you know, so um, these size houses were meant to be run with help. One person could not do it, especially if that person also wanted to have a life outside the house (...)

The creative aspect of home-making, beyond the more drudging chores of cleaning and tidying, involves making meals for friends and family, and generally creating a cosy house. The husbands and children living in the house, and presumably the guests as well, are all entitled to this feeling of cosiness. Friends seem to also be a source of pressure, when it comes to maintaining certain standards. As Selma explains, "it's amazing how women judge women."

Social pressure also leads to the acquisition of new appliances, as stated earlier around the example of the Thanksgiving meal which prompted the acquisition of a new refrigerator, as the hostess had expressed feelings of shame at having to store food outdoors. In another example, a woman explains her purchase of an electric water boiler and a juicer in these terms:

> Um, I recently bought a juicer in the last two years because it's been the rage, everyone is juicing, detox, you're juicing, you don't have a juicer, you need to buy a juicer, okay, let's go buy a juicer. The kettle... even my kettle is recent... because I always had the kettle that you put on the stove (...) And friends would come over and they'd be like, jeez, how long does it take for you to boil a cup of tea for me.

The Nutri Bullet, a blender that promotes the healthy extraction of nutrients from fruits and vegetables, is a trending appliance in these homes and came up in several interviews. In Esther's house, the newly purchased blender was sitting in its box, ready to be used for the first time. The health benefits of the machine were the primary reason for purchase, based on hearing about the Nutri Bullet from friends. Friends can play an important role in appliance acquisition.

For mothers of teenage children in this group, some felt that younger generations are increasingly expressing feelings of entitlement towards a privileged social position, reinforced through their peer groups. This demonstrated, for Rada, a form of excess. She explained how a new social network had been created:

> Cool Kids with Cash, and it's an Instagram account on them, and they won't let you access it, and these kids are posting pictures of Hermès, Gucci, Louboutin, they're spending a thousand francs a day, and I am thinking, where is this money coming from? (…) I mean, my daughter says to me, oh, I am going to La Reserve [a five-star hotel] for coffee, I am like, really, because the coffee shop down the road is not good enough?

What is normal is therefore reinforced by social position across generations, creating expectations around normality when it comes to consumption practices. We now turn to how normality changes in different contexts of consumption.

Lock-In and the Un-locking of Normality Across Different Contexts and Cultures

One of the most striking aspects of this group is that they are well travelled, have lived in different countries and contexts, such as developing countries or mega cities, where they have picked up different habits, and seem to be incredibly adaptable to changes in consumption that might result from being in these different spaces—at least for limited amounts of time. Out of the ten women interviewed, six had second and sometimes third homes, including winter chalets in the nearby Alps, a summer house off-the-grid in Spain, or homes in other places for vacations.

Sally remembers living in New York City in the 1980s, when air-conditioning was less popular than today and, as she recalls travelling to work by subway in the summer months, when body odour was something quite normal. Sally also owns a second home in the Spanish countryside: it's off the national grid and they have gone to great pains to install renewable energies and change their consumption patterns, involving sun-cookers, less showers and less clothes washing. Sally's daughter is more careful about washing clothes in Spain, but falls back into her regular habits when back in Switzerland. For example, a t-shirt worn once will be washed, as opposed to being worn several times before washing.

Nathalie grew up in South Africa and based on that experience she always hangs her clothes to dry, considering it too much of a "luxury" to dry them by machine, as she puts it. Selma grew up in Amman, Jordan, and as a result is very conscientious around water consumption in the home—as she recalls living in situations of water limits, with clear restrictions on when gardens could be watered. Rada experienced frequent power outages when visiting India in her youth and also recalls boiling kettles for washing up in the United Kingdom. Esther also recalls growing up in the United Kingdom, when blackouts were common. I asked her how she lived in those moment, she explains: "Fine, we just adapted. You were told that, that there would be no, there would be no electricity between seven and nine that night and so there wasn't." Habits may have been formed at early life stages that are carried forward in different contexts.

In some cases, habits do not change across contexts: Melanie and her family, for example, enjoy a Jacuzzi in each of her homes, from the Swiss Alps, to Geneva, to Vermont. Certain practices are very much maintained through this material lock-in and family expectations around a good life, leading to what can be termed an "expat lifestyle" carried forward in different contexts around the world. Yet in Vermont, Melanie explains that social expectations are very different from those she experiences in Switzerland. Her Vermont cabin is a space where standards can be relaxed, where gatherings are much more casual and where expectations around tidiness are lowered. The Geneva expat community gives more importance to keeping up appearances and material forms of consumption, whereas in Vermont forms of downshifting towards a simpler mode of living are recognized as desirable. The meaning of "status" therefore can change in different contexts, within a similar social group.

To summarize, while habits can be formed in different contexts, during childhood for example, and carried into new contexts, some practices are very much maintained across different contexts—leading to a standardized expatriate lifestyle. There do seem to be opportunities to "un-lock" consumption from material and social dimensions, however, in that certain standards are changed—related to electricity distribution and availability, for the first dimension, and social expectations around tidiness and appearances, for the latter. There is also the question of temporality, as holidays represent a more limited period of time when expectations can be shifted and reduced, without that necessarily leading to the reconsideration of practices when back home.

3.4   Conclusions

This chapter has sought to explore some of the underpinning dimensions of energy-using practices among an affluent social group in Geneva, or people who self-identify as being part of the expatriate population. With high financial, social and cultural capital, the people interviewed for this research have large and multiple homes, abundant appliances, and travel frequently for professional reasons and leisure. The multiplying of appliances and powered devices can be seen as a worrisome trend whereby multiple refrigerators and dishwashers and other household technologies might become a new standard. While their own experiences elsewhere may lead to a slackening of the material dimension of consumption and social expectations around upholding certain norms, such as less tidiness in a Vermont home or appliance usage in an off-grid Spanish villa, the issue remains that of *overall* energy consumption. In the case of second and third homes, it is not just a multiplying of appliances but a multiplying of all living space with the associated energy consumption for heating and cooling these homes.

In relation to notions of excess, there is no explicit upper limit to consumption that is enunciated in discourse, but rather a sense that excessive consumption takes place beyond what this social group considers as normal—or an imagined boundary. People within the same social group do not necessarily question their own consumption practices, even with the doubling of appliances or living spaces. One explanation may lie in the visibility of other consumers beyond that imagined boundary, who always appear to consume far more, and exemplified in media coverage around jet-setting celebrities in "McMansions". Normality is therefore socially constructed in opposition to expectations around what is put forward as "excessive" in society. Social lock-in therefore comes into play in two ways: people are locked into social standards, such as maintaining order, cleanliness and cosiness, towards family and friends; they can also be locked into societal expectations, as communicated through the general media. Material lock-in is also apparent, in that people can be locked into their appliances and home features, such as a blender, a lift or a swimming pool, but they can also acquire new appliances in relation to the size of their homes – larger homes enable the acquisition of new or multiple devices. There is also a strong interplay between these two forms of lock-in: the material arrangements of a home bring convenience and comfort, or are forms of display to others, tying in with social expectations.

Based on this analysis, the opportunities for destabilizing practices and challenging expectations around consumption could take into account three factors when considering this particular group of people. First, the opportunity to capitalize on how practices play out in different contexts and the significance of demonstration sites. The women interviewed for this chapter all have experiences of living with much less stuff and in smaller spaces, for example in settings where consumption is restricted, such as the off-grid summer home in Spain. High-quality smaller homes could be developed as demonstration sites where new practices can be experimented with, particularly by people with the financial and social capital to do so. Second, the value of time: Jacuzzis need to be maintained, appliances fixed or replaced, and there is quite a bit of effort and time going into the upkeep of these objects. For some people this is a source of frustration, despite the availability of domestic help which also entails time being spent on their management. If there is a sense that too much stuff takes up too much time, this could be explored as a way of promoting forms of downshifting. Slowing down time, taking your time, is something that people in this group may well look for—being rushed and organizing large households creates feelings of fatigue and stress for many of the women interviewed.

Third, throughout all of the interviews, the significance of the social realm was highly apparent. The advice and recommendations of friends, what other people are doing in their social group, are all significant. There would be a need for trendsetters, within this social group, to take the lead in engaging with less energy-intensive forms of consumption. If social groups can be a form of pressure towards appliance acquisition, they can also be a source of inspiration: if a trend towards reduced energy consumption were ever to catch on, it would not be thanks to a utility company campaign for this social group, but because friends and family are directly involved in the sharing of new, energy reducing practices. Social networks are therefore both a medium for maintaining the status quo, as well as a possible vector towards un-locking practices towards sustainable consumption transformations. That being said, social lock-in also involves the role of media practices, and how these channels communicate around normality and setting expectations around what is excessive. Normality is therefore constructed within specific social groups and cultural settings, in relation to the material and social dimension of consumption, and involving appliances, space availability, the influence of peers, but also the general media.

With that said, the author cannot help but reflect on her own consumption standards, which may be below that of a group of expats, but are far above those of the many millions worldwide who lack access to reliable energy. Yet my own consumption practices are, within my social group, understood to be normal. More research is needed on understanding expectations, defining excess and contesting normality across all socio-economic groups, particularly in the so-called developed world.

Acknowledgements This chapter benefited from the careful review of the editors, and I greatly appreciate their constructive feedback and comments. Many thanks to my colleague Béatrice Bertho at the University of Lausanne for her precious collaboration on research design, implementation and analysis. I also thank all the women who graciously agreed to be interviewed for this study and opened their homes to me. The Swiss National Science Foundation is gratefully acknowledged for funding this project, under the National Research Program on Managing Energy Transitions (NRP71), coordinated by Suren Erkman at the University of Lausanne and co-coordinated by Marlyne Sahakian, at the University of Geneva as of August 2017. Much of the research on which this paper was based was undertaken during her time at the University of Lausanne.

Notes

1. The word "expatriate" is derived from the Latin words *ex*, out of, and *patria*, country or fatherland. The term "expat" is usually in reference to white collar, professional workers living outside of their countries of origin.
2. It is not uncommon for expats in Geneva to rent rather than own their homes, as rental fees are often covered by the companies where they are employed. They pay for their own utility bills, however. In rented Swiss homes, refrigerators are sometimes included in the kitchen, although not always.
3. As part of the larger research project in Western Switzerland, multiple refrigerators and freezers were also found among households in lower socio-economic groups for financial and social reasons: for the purpose of buying food more economically in bulk and sharing meals with large and extended families, in the case of a migrant family from Kinshasa, the Congo, living in Geneva.

Bibliography

Anantharaman, M. 2016. Elite and ethical: The defensive distinctions of middle-class bicycling in Bangalore, India. *Journal of Consumer Culture* 1–23. First published 8 March. doi:10.1177/1469540516634412.

Arthur, W.B. 1989. Competing technologies, increasing returns, and lock-in by historical events. *The Economic Journal* 99: 116–131.

Bourdieu, P. 1979. *La distinction. Critique sociale du jugement*. Paris: éditions de Minuit.

Bourdieu, P., and L.J.D. Wacquant. 1992. *Réponses: Pour une anthropologie réflexive*. Paris: Seuil.

Bruijnis, M.R.N., V. Blok, E.N. Stassen, et al. 2015. Moral "lock-in" in responsible innovation: The ethical and social aspects of killing day-old chicks and its alternatives. *Journal of Agricultural and Environmental Ethics* 28: 939–960.

Dubuisson-Quellier, S., and M. Plessz. 2013. La théorie des pratiques: Quels apports pour l'étude sociologique de la consommation? *Sociologie* 4: 451–469.

Flam, H., and J. Kleres, eds. 2015. *Methods of exploring emotions.* London: Routledge.

Hitchings, R., and S.J. Lee. 2008. Air conditioning and the material culture of routine human encasement: The case of young people in contemporary Singapore. *Journal of Material Culture* 13: 251–265.

Hochschild, A.R. 1979. Emotion work, feeling rules, and social structure. *American Journal of Sociology* 85: 551–575.

Perkins, R. 2003. *Technological "lock-in".* International Society for Ecological Economics. Available at: http://www.isecoeco.org/pdf/techlkin.pdf

Plessz, M., S. Dubuisson-Quellier, S. Gojard, et al. 2016. How consumption prescriptions affect food practices: Assessing the roles of household resources and life-course events. *Journal of Consumer Culture* 16: 101–123.

Rip, A., and R. Kemp. 1998. Technological change. In *Human choice and climate change*, ed. S. Rayner and E.L. Malone, 327–399. Columbus: Battelle Press.

Sahakian, M. 2015. Getting emotional: Historic and current changes in food consumption practices viewed through the lens of cultural theories. In *Putting sustainability into practice: Advances and applications of social practice theories*, ed. E. Huddart Kennedy, M.J. Cohen, and N. Krogman, 134–156. Cheltenham: Edward Elgar.

Sahakian, M., and H. Wilhite. 2014. Making practice theory practicable: Towards more sustainable forms of consumption. *Journal of Consumer Culture* 14: 25–44.

Schatzki, T.R. 1996. *Social practices: A Wittgensteinian approach to human activity and the social.* New York: Cambridge University Press.

———. 2002. *The site of the social: A philosophical account of the constitution of social life and change.* Pennsylvania: Pennsylvania State University Press.

Shove, E. 2003. *Comfort, cleanliness and convenience: The social organization of normality.* Oxford: Berg.

Spaargaren, G. 2011. Theories of practices: Agency, technology, and culture, exploring the relevance of practice theories for the governance of sustainable consumption practices in the new world-order. *Global Environmental Change* 21: 813–822.

van Griethuysen, P. 2010. Why are we growth-addicted? The hard way towards degrowth in the involutionary western development path. *Journal of Cleaner Production* 18: 590–595.

Warde, A. 2004. Theories of practice as an approach to consumption. *Cultures of consumption programme* (working paper 6).

Marlyne Sahakian is an Assistant Professor in Sociology at the University of Geneva, Switzerland. Her research interest is in understanding consumption practices in relation to environmental impacts and social equity, and identifying transformative opportunities towards more sustainable societies. She is currently coordinating Swiss and European research projects on household energy and food consumption. Her recent work includes *Keeping Cool in Southeast Asia: energy consumption and urban air-conditioning* (Palgrave Macmillan, 2014) and an edited volume *Food Consumption in the City: Practices and patterns in urban Asia and the Pacific* (Routledge, 2016). She is a founding member of Sustainable Consumption Research and Action Initiative (SCORAI) Europe.

4

Understanding Temporariness Beyond the Temporal: Greenfield and Urban Music Festivals and Their Energy Use Implications

Michael E.P. Allen

4.1 Introduction

Organised events are an integral feature of the ordering of society and social life. Taking many different forms, organised events provide a focus for the shared performance of practices, in time and usually also in space, by those involved. As such, they also contribute to the ongoing patterning of energy demand in time and space, but in ways that are very much dependent upon the characteristics exhibited by particular events. This chapter explores specifically what a focus upon the temporariness of organised events can contribute to understandings of energy demand.

Temporariness is an obvious, but under-theorised, feature of organised events. Where temporariness is theorised it is usually understood in terms of duration. Getz (2005: 15–16) acknowledges the temporary nature of events, seeing it as their defining characteristic. Indeed, Getz's definition of planned events as 'temporary occurrences with a predetermined beginning and end' (2005: 16), foregrounds their duration. Whilst Getz also

M.E.P. Allen (✉)
DEMAND Centre, Lancaster University, Lancaster, UK

© The Author(s) 2018
A. Hui et al. (eds.), *Demanding Energy*, DOI 10.1007/978-3-319-61991-0_4

acknowledges that events may recur at regular intervals, his emphasis upon duration provides limited scope for examining the implications of temporariness upon the organisation and energy use of events.

This chapter argues that expanding understandings of temporariness to include both temporal (duration, rate of recurrence) and spatial (location, infrastructural arrangements) characteristics provides a more robust basis for considering the implications of temporariness upon energy demand. By applying these characteristics to the study of greenfield and urban music festivals, the chapter demonstrates how temporariness can contribute to understandings of energy demand in time and space, not only in relation to music festivals, but for a wide variety of other organised events, from local markets to international sporting tournaments.

Working with temporariness in this way makes three contributions. First, it demonstrates that the conceptualisation of temporariness itself is complex, requiring careful consideration of multiple temporal and spatial characteristics by the researcher. Second, it draws attention to the difficulties in identifying what is and what is not a part of any given organised event, thereby problematising the bounding of the event, and accounting for its energy demand implications, in space and time. Finally, it highlights that understandings of temporariness in the literature dealing with energy use and organised events can produce a restricted focus on the event itself. Failing to unpack the ways in which organised events are disruptions to regularised routines and practices can lead to an incomplete representation of the place of events within the overall picture of energy demand (e.g. Bottrill et al. 2007; Johnson 2015: 2).

The chapter proceeds as follows. First, drawing upon social practice theory, a way of understanding organised events is outlined. The examples of greenfield and urban music festivals are then introduced as a means by which to explore an expanded conceptualisation of temporariness for the study of organised events and energy demand. Following this, the aforementioned temporal and spatial features are considered in turn. It is concluded that consideration of temporariness can help foster a deeper understanding of the energy use of organised events.

4.2   Organised Events and Temporariness

Before looking at why temporariness is a useful concept, it is necessary to understand what is meant by an organised event. The term 'organised event' is a high-level categorisation similar to Getz's concept of the 'planned event' (2005: 15–16) that includes all events that are deliberately conceived and professionally managed. Organised events, such as music festivals or football matches, are conceptualised here as bundles of practices that co-locate in time and, usually, also in space (Shove et al. 2012). Practices are understood as 'temporally dispersed nexus[es] of doings and sayings' (Schatzki 1996: 89) and they form a bundle when they temporarily establish 'loose-knit patterns based on … co-location and co-existence' (Shove et al. 2012: 81). The bundle of practices in a football match can thus include practices that take place at the event, such as playing football, refereeing, cooking, serving food and stewarding. In addition, going beyond the understanding of a practice bundle as being co-located, temporally and spatially distributed practices are also considered as part of a bundle when they play an important role in generating and pre-figuring the configuration of practices and opportunities for energy use at organised events. At a football match, such practices include travel practices of teams and spectators, and organising and administrative practices that make the occurrence of the match possible and affect the configuration of the bundle.

Organised events can be described as temporary comings together of social practices outside the framework of regularised, mundane and routine forms of everyday life. They exhibit a huge degree of variation, with different scales, primary purposes, durations, rates of recurrence and, importantly, locations and infrastructural arrangements. For example, league football matches last around two hours, recur every week during autumn, winter and spring and draw largely upon permanent infrastructural arrangements. In contrast, the Olympic Games recur only once every four years, involve multiple constituent events that take place each day over a two-week period and require much in the way of new infrastructure, such as sports arenas and an Olympic Village. Again, in contrast, a local market may last from 9 am to 4 pm, occur monthly and utilise a mixture of makeshift and permanent infrastructure in the form of market stalls, a town square and street lighting. A travelling fair may last 12 hours a day over a two-week period but only during certain seasons; it recurs perhaps once a year in any given location and relies upon makeshift infrastructural arrangements. These examples of variability in the characteristics of different types of events demonstrate that being temporary is not a uniform condition.

The literatures both on practice theory and events management have tended to overlook or not fully consider this variation and multiplicity within temporariness. There are examples of organised events being studied from a practice theory perspective (e.g. Schatzki 2010; Blue 2013) but practice theory mainly focuses on regular and routinised practices at home (Shove 2004; Jalas and Rinkinen 2016), in the workplace (Yli-Kauhaluoma et al. 2013) and on the move (Hui 2012; Cass and Faulconbridge 2016). Organised events fall outside the pattern of everyday activities: they change how social practices such as eating and socialising are performed and cluster together. This, in turn, impacts the energy demand of organised events, as 'energy is used not for its own sake but as part of, and in the course of, accomplishing social practices' (Shove et al. 2014: 7). This is particularly important when considering leisure activities, such as sporting and live music events.

Greenfield music festivals, which are characterised by the temporary coming together of social practices in space and time, serve as a good case for examining temporariness. A number of characteristics have coevolved throughout the history of greenfield music festivals, including duration (usually a weekend in summer), rate of recurrence (usually annual), location (usually pastoral) and degree of infrastructural impermanence. As the discussion below demonstrates, examination of the relationships between these characteristics helps us to understand the influence of temporariness on the types of social practices that make up festivals and how they are performed.

Urban music festivals are likewise characterised by the temporary coming together of social practices in space and time. The characteristics of temporariness found in urban music festivals, however, provide a counterpoint to greenfield music festivals that accentuates the importance of looking beyond duration. Like greenfield music festivals, many are multi-day and annual, but they differ in terms of their location and materiality, relying more on fixed infrastructural and material arrangements than makeshift ones and, as will be discussed, often enrolling fewer and less diverse practices into their performance.

Examining the differences between greenfield and urban music festivals in more detail can therefore help us to understand how practices temporarily bundle together and how these practices change when they become part of events. It can also provide insight into flexibility and variability in practices. In what follows I draw on information gathered from three sources: insights gained through attending and observing two greenfield and two urban festivals; a programme of interviews with people involved in varied aspects of organising and provisioning festival events, including festival organisers, sound engineers, festival workers and service providers; and historical material from the secondary literature on UK music festivals.

4.3 Characteristics of Temporariness

In this section, the key features of temporariness are described and explored in turn: temporal (duration, rate of recurrence) and spatial (location, infrastructural arrangements). These features are important both in their own right and also, specifically, due to the energy demand implications of live music festivals.

Temporal Features

Getz's emphasis upon duration is not misplaced, as the duration of events affects the configuration of the social practice bundle, as well as individual practices and how they are performed. Events lasting many days or weeks can be expected to involve a wider array of practices and greater number of different practices than events that take place over a few hours. Yet how long an event lasts is a matter of interpretation. Greenfield festivals are usually open to the public for at least three days, and sometimes longer. The different types of practices that are part of the festival bundle, however, may be shorter or longer in duration. For example, entertainment practices may last two or three days, whilst campers may be allowed to stay for an extra day before or after the entertainment. The process of building and deconstructing the festival site, by contrast, may last months in the case of bigger festivals.

There are also variations in terms of the continuity of participation in events. When a greenfield festival is open to the public, its duration tends to be unbroken because festival-goers stay on the festival site. This leads analytically to additional practices being included in the festival bundle, such as camping, which may include such practices as cooking, eating, washing and sleeping. Such practices are therefore included as part of the festival bundle due to the continuous duration of the event.

Multi-day urban festivals, by contrast, cannot be deemed to feature practices related to camping, such as sleeping and washing, due to the lack of continuous participation. However, this does not mean that such practices do not take place during urban festivals. Entertainment practices may take place over the course of three days (or even longer) at an urban festival, whilst setting up and taking down are usually accomplished in a single day either side, rather than sharing the protracted experience of greenfield music festivals. However, when the entertainment practices are not continued overnight, this represents a break in the continuous operation of the festival, limiting the number of practices that take place on the festival site. The practices that are associated with camping at greenfield festivals (cooking, eating, washing, sleeping, etc.) tend to take place in the periods of hiatus during an urban festival, and outside the remit of its organisation, at people's homes and in hotel rooms and restaurants.

This issue of the temporal bounding of urban festivals highlights the ambiguity that surrounds the attribution of practices to urban festivals. Many of the practices that occur at greenfield music festivals because of the duration of the festival, such as washing or showering, also occur during the course of an urban festival but, as they occur off-site and are not organised by festival managers, they are difficult to monitor and make little sense to attribute to the urban festival.

This has obvious implications for the perception of energy use and energy accounting at urban and greenfield music festivals. For example, while I have argued that it is not part of an urban festival per se, washing (showering or bathing) is more likely to take place during an urban festival than a greenfield music festival and more likely to use energy in its performance. This is significant for energy accounting as, at greenfield festivals, practices such as showering draw energy on-site, from generators, gas bottles or solar systems. As such the energy use from this practice forms an unquestioned part of greenfield festivals' energy use. The status of showering as a part of the urban festival practice bundle, however, remains ambiguous due to the temporal and spatial discontinuity of the event. Since festival-goers' showering during urban festivals takes place off-site and draws energy from sources other than those provided by the festivals, the energy use associated with showering, I argue, should not be attributed to the urban festival. Yet, if we consider that greenfield festivals often involve a decrease in showering practices amongst festival-goers during the festival, relative to their everyday lives outside the festival, whilst showering during urban festivals likely remains at the same frequency as in the festival-goers' everyday lives, we see how a focus on the temporal aspects of different types of events reveals that there is more to the story of energy use at music festivals than simply how much energy is used during the festival's official hours of operation. The break in continuity that occurs in urban festivals puts these everyday life practices beyond the influence of festival organisers and speaks to the difference between the everyday energy use of social practices and the energy use of equivalent practices at greenfield music festivals.

Duration is also important when considering how long practices at the festival are using energy in comparison to other everyday life practices that are foregone whilst the festival is taking place. Greenfield festivals continue 24 hours a day for those festival-goers who stay on-site, who make up the majority at such festivals. Provision, therefore, needs to be made for the practices of eating, brushing teeth, washing and sleeping. Historically, provision of basic services was an issue at some festivals. Inadequate provision of clean water points, open trench toilet facilities, long queues for food and food shortages were features of some festivals in the 1960s and 1970s (Advisory Committee on Pop Festivals 1973; Sandford and Reid 1974; Hinton 1995). The provision of additional facilities in recent years, like showers and mobile phone charging, indicates a trend towards the inclusion of more energy using practices, and is linked to the continuous multi-day duration of the festivals. Practices such as showering and charging mobile phones would not necessarily be expected at events of a shorter duration, such as a football match.

Lighting has become a significant part of the 24-hour provisioning at greenfield music festivals, being required for the whole of the festival site during the hours of darkness, including car parking areas. However, lighting was once not a ubiquitous feature of festivals: many were criticised for being unsafe due to lack of lighting provision (Advisory Committee on Pop Festivals 1973). As festival organisation and regulation have developed, the degree of light provisioning, which helps facilitate the everyday lives of festivals-goers, has increased markedly. Whilst technologies, such as LED lighting, have been instrumental in reducing the costs of lighting, and thereby energy, at festivals, it is clear that moves towards increased lighting provision have energy implications for festivals.

Despite increased provision for an increasing number of practices at greenfield festivals, 100,000 people living in a field for a weekend is a very different prospect from 100,000 living in houses over the same period. Camping and its associated practices have already been mentioned as examples of this, but it is also the case that some high energy using practices, such as driving, are discontinued for the duration of a greenfield festival. Driving is a focus of much of the literature on the energy use of greenfield festivals, as it makes up between 67 percent and 80 percent of the total CO_2 emissions (Bottrill et al. 2007: 49; Johnson 2015: 2). Whilst this is a significant issue for greenfield music festivals, it does not take into account the number of miles that were not driven as a result of drivers being at a festival. Furthermore, this speaks more to the prevalence of high energy using mobility practices in our society than to an intrinsic problem with greenfield music festivals. Equally, the provision of mobile phone charging facilities is a result of changes in everyday life practices in society at large, making it unfair to criticise greenfield music festivals for offering this service. The changing energy provisioning and use at festivals therefore needs to be considered alongside similar changes in everyday life outside festivals.

The links between the duration of practices and their energy-intensive provisioning is not limited to everyday life practices. Music practices have extended in their duration over the course of a day at greenfield music festivals from a mere four and a half hours in the 1950s (Montagu 2000) to, in some cases, 24 hours. Today, it is common for festivals to offer stages in arenas that open from around 11 am to 11 pm and often further music from DJs until the early hours of the morning. Interviews with sound engineers revealed that the power rating and the amount of equipment needed for each stage is largely determined by the physical requirements of the job; by the size of the audience, the setting (a tented structure or open air), and the duration of use. Contemporary expectations about the equipment enrolled in sound engineering and stage provisioning practices (e.g. the digitisation of equipment, the types of speakers used and the presence of big screens) also influence the materiality of practices and ultimately their energy use. Despite the influence of contemporary expectations about equipment on energy use, the need for this energy use arises due to the aforementioned physical requirement of delivering clear, quality sound to a given number of people. As long as the requirement to communicate music to large numbers of people at a given stage remains, this means that the amount of energy used for each stage is greatly increased by the number of hours for which it is operational. Therefore, from a durational standpoint, energy savings could be made by reducing the number of hours that a stage is operational. Reducing the number of hours a stage is operational (along with reducing the number of stages) is also a way of ensuring that energy is used more efficiently in terms of the number of people being entertained at a given time. This is because stages usually attract fewer festival-goers earlier in the day. Such moves, however, would be hampered by festival organisers' perceptions that having a greater number of acts at a festival will increase its appeal, which necessitates an increase in the number and absolute duration of music performances. Indeed, this increased duration of music performances may have become normalised in festival-goers' expectations.

As has been discussed, duration is important to understanding the temporariness of organised events because of its varied implications for different practices within the bundle, its relationship with continuous and discontinuous participation or operation, and its links to how festivals are provisioned for various everyday and music practices. Another temporal feature of live music events that warrants further consideration, however, is rate of recurrence.

As mentioned above, different events have different rates of recurrence, be they weekly league football matches, monthly markets or quadrennial international sporting tournaments. This has an impact on the practice bundles that make up the events and the wider set of practices that are required to facilitate them, such as the building of stadiums. Rate of recurrence is of particular importance in material terms and so becomes closely linked to spatial features of temporariness. This link will be explored further in the next section.

Rate of recurrence, however, also has important ties to the development and evolution of festivals. Historically, summer was a quiet time for live music events in the UK. Both greenfield and urban music festivals took advantage of the otherwise quieter summer season to stage performances (McKay 2005). Greenfield music festivals in particular have exploited this summer niche, growing dramatically in number in the twenty-first century (Anderton 2008). McKay (1996: 12) argues that greenfield music festivals are the successors to 'seasonal and nomadic festivals and fairs in the country and the city', which themselves are romanticised reproductions of early events. The Beaulieu Jazz Festivals (1956–1961), arguably the earliest examples of greenfield music festivals in the UK, are seen as a successor to the Beaulieu Fairs of the nineteenth century (McKay 2005). The origin of music festivals in these yearly celebrations partly explains why individual music festivals tend to take place annually.

As greenfield festivals tend to occur a year apart at any given site, organisational practices follow a temporal pattern. First come practices associated with preparing the festival, then practices associated with running the festival and, finally, practices associated with striking (taking down) the festival. Given the relatively short duration of greenfield festivals, this sequence does not involve practices associated with the long-term occupation of land or the frequent recurrence of events. Such practices would bring with them different material arrangements and would change both the configuration of the festival bundle and the patterns of energy use. The practices that make up the bundle do not, for example, include practices focused on keeping the site in such condition as to enable weekly festivals. This is particularly important for festivals that have experienced heavy rain that turned the ground to mud. Given the scale of the clean-up operations following larger festivals, it is easy to see that practices during the festival would need to change in order to ensure the site was habitable week after week or month after month, to avoid serious environmental and health problems.

In order to explore the implications of these issues, and the rate of recurrence of festivals, upon energy use, further attention will now be given to the spatial features of temporariness.

Spatial Features

While temporariness might be commonly understood as solely a temporal phenomenon, this chapter argues that in order to consider the energy implications of organised events, their spatial characteristics must be investigated in equal detail. In terms of the cases examined here, the location of music festivals is a key feature distinguishing greenfield and urban festivals. How festival sites are spatially bounded, the regular land use of the sites occupied temporarily by festivals, and the affordances of different locations are all spatial characteristics that shape understandings of their temporariness. Location also influences the infrastructural arrangements that support music festivals. Infrastructural arrangements may have a greater or lesser degree of permanence: this is an important part of temporariness. This section explores the dynamics of these spatial features and how this impacts energy use.

The effect of durational continuity and discontinuity on the configuration of festival practice bundles is outlined above. However, there is an additional spatial element to this. Greenfield festival sites are spatially continuous and arranged to incorporate a full range of practices, from music and entertainment practices to everyday life practices. In contrast, the sites of urban festivals are often discontinuous, being formed of discrete venues spread across a pre-existing, public, urban area. As such, urban festival sites do not incorporate full-scale provision for everyday life. The lack of on-site provision for everyday life practices at urban festivals means that many practices take place off-site. This has already been shown to be the case for practices like sleeping and showering that can be expected to take place during the festivals' overnight hiatus. However, the geography of urban festival sites can mean that practices such as eating also take place off-site during the festivals' hours of operation.

The lack of a single space that can be characterised as a homogenous and spatially continuous festival site adds to the ambiguity around exactly which practices are a part of the urban festival bundle. The utilisation of public spaces by many urban festivals also makes it difficult to identify exactly who is and who is not a festival-goer. Festivals that are free of charge or take place in a series of public spaces where access is not normally restricted can lack a clearly defined inside and outside. This presents difficulties in determining exactly who should be classified as a festival-goer. This ambiguity makes it difficult to be sure exactly what and whose practices are being performed as part of the festival and how, thus blurring the relationship of certain practices to the festival and diminishing the role and responsibility of the festival organisers for them.

Furthermore, it must be remembered that greenfield festival sites do not draw on energy for year round operation and maintenance in the same way that urban festivals do. A field does not have a base load in the way that a building does (consider the energy used by a tent against that of a building). Organisers of the Glastonbury Festival, which recurs annually at Worthy Farm in Somerset (with the exception of a fallow year every five years in which no festival takes place), point to concrete examples of the positive environmental impact of annual festivals. They point out that the annual rate of recurrence allows the area of the festival site to be managed in such a way as to promote biodiversity and has led to the cessation of resource and energy-intensive farming methods (Bowdin et al. 2011). Glastonbury may well be exceptional in this regard; however, the point that most festival sites lie relatively dormant in terms of energy use for over 11 months of the year is significant, as it contextualises the importance of temporariness for understanding organised events and their use of energy.

In contrast, urban festivals tend to take place on sites that are used regularly, often for similar purposes, such as city halls, purpose-built concert venues or pubs. The temporary occupation of a site by a music festival may lead to an increase in the energy expenditure normally associated with the site, though in some cases this may only be marginally so. It is likely, however, that the difference between the business as usual energy expenditure and the festival-related energy expenditure of the site will be closer for urban than for greenfield festivals. This is not to argue that either greenfield or urban festivals are better in this regard. Rather, it is to point to the events as temporary departures from the festival sites' regular use and energy expenditure to a greater or lesser degree, for better or worse. Again, this could say more about the amount of energy regularly used at the site than about the festival.

The differences between rural and urban locations also affect the provisioning of festival sites. Urban festivals do not need to provide services for the everyday lives of festival-goers due to the affordances of urban areas. Public transport systems, hotels, restaurants and local shops provide for the needs of festival-goers in urban areas, making it unnecessary for festival organisers to do so. Where provision is made for the everyday life practices of festival-goers by festival organisers, it comes in two forms: pre-existing facilities in the venues and pop-up shops. The utilisation of pop-up shops adds to the degree of impermanence of urban festivals. By contrast, the spatial continuity of greenfield festival sites means that many everyday practices are catered for on site. The increase in the number of everyday life practices that are supported at the festival, coupled with the lack of permanent infrastructural support for such practices at the festival site, leads to a greater reliance on makeshift arrangements at greenfield festivals relative to urban festivals.

As this highlights, the infrastructures that support music festivals and other events, and their temporality, are crucial for understanding an event's temporariness. The degree of infrastructural (im)permanence that supports social practices at an event is significant for the energy make-up and provisioning of events and can influence both the kinds of social practices and their performance. Urban festivals utilise existing infrastructures and materials, including the energy infrastructure and building stock, to support events. In this respect, urban festival infrastructure exhibits a greater degree of permanence than greenfield festivals, which have to build a temporary settlement, complete with power infrastructure, and entertainment and living areas. A multipurpose venue, such as an arena, may require building a stage for each event but not building the venue itself or the energy infrastructure that supports the performances. Even outdoor stages at urban festivals can be powered using the existing energy infrastructure of neighbouring buildings. This means that the energy using practices found annually at greenfield festivals, associated with putting together and taking down what amounts, in some cases, to a temporary city, are not present in the urban festival bundle.

Different types of practices are affected by the independence of greenfield festivals from permanent infrastructure. First, several practices come into being at festivals that would not be necessary at events taking place in permanent structures; for example, constructing or bringing in temporary substitute structures (tents, marquees, toilet blocks, walk ways, lighting, car parks, etc.). Second, practices associated with music, such as sound engineering and making and listening to music, can continue in some respects as they would at any music event. Much of the equipment used is the same and the necessary power is provided to make it work. In contrast, the performance of practices related to the everyday lives of festival-goers, such as cooking, washing and keeping warm, can exhibit a great deal of discontinuity inside and outside the festivals, largely due to lack of connection with permanent infrastructure. These practices are changed by the temporary infrastructural arrangements of festivals; for example, the provision of food not only requires the transport of food to the site but also the means to keep it fresh, cook and distribute it. At festivals, this is usually provided by mobile food vendors in motorised vehicles or by members of the audience using cool boxes, fires and gas stoves.

Greenfield music festivals rely mainly upon makeshift infrastructure (e.g. tents), although there are instances where permanent infrastructure, such as transport infrastructure, is drawn upon. Putting together and taking down infrastructural arrangements can be very energy intensive and is an expenditure of energy not required by events linked to more permanent infrastructures. On greenfield sites, the lack of building infrastructure is compensated for by tented structures and modular stage setups. It takes a number of weeks (even months) prior to the opening of a festival to assemble the infrastructure at the site, reflecting the location and its lack of preparedness for this type of event. The link between the temporary nature of the event and the impermanence of the infrastructure is captured by Glastonbury Festival's environmental policy, which describes the festival as 'a one-off event in a field' (Bowdin et al. 2011: 179).

The lack of permanent services and infrastructure, such as waste collection and a sewerage system, create issues for festival organisers, but the influence of these factors on energy use relative to permanently maintained infrastructures has not been fully explored in the literature. It suffices here to mention that the lack of permanent waste water and sewerage infrastructure provides both constraints and enablements in terms of greenfield festival organisers' ability to influence energy use. This is illustrated by Glastonbury Festival's 2008 environmental policy, which states the aim of reducing the CO^2 emissions (and, by proxy, energy use) associated with the collection of waste water and sewage by finding ways to process waste on site. This points to the greater malleability of temporary infrastructures as compared with permanent infrastructures, which have a higher degree of obduracy (Hommels 2005).

Makeshift infrastructures also provide an extra degree of flexibility in terms of energy choice. The separation of greenfield festivals from the country's permanent energy infrastructure provides both constraints and enablements in terms of the energy use of practices. The use of off-grid power generators to provide energy, commonly powered by diesel, biofuel or solar, means that fuel sources are not dictated by the energy composition of the national grid. The use of generators also means that total energy use at greenfield festivals must be predicted in advance to prevent power loss and economic loss from the use of inappropriately sized generators (Johnson 2015).

For urban festivals, the infrastructure is already in place to support the festival, making it possible to simply overlay a music festival in an urban area with only minimal disruption to the everyday operation of these venues. Fewer additional practices or arrangements are therefore needed for urban compared with greenfield festivals. This means that festivals are able to save energy relating to practices of provisioning the festival site, but also provides limitations in the form of being tied to the energy make-up of the national grid. It may also mean that organisers are less conscious of the energy use of the festivals as it does not need to be worked out in advance and is usually paid for by the venues and perhaps only indirectly charged to the festivals in terms of hiring fees.

While the temporality and nature of infrastructures themselves is important for understanding the energy implications of festivals, they also need to be considered in relation to everyday practices. That is, it is necessary to think of the festival as a break in routine when assessing the energy use of festivals. This is particularly important for greenfield festivals. Energy use can only be fully understood by situating practices at the festival within the broader context of everyday social practices and comparing the social practices of the everyday lives of festival-goers with their day-to-day use of the festival site. For example, washing may be foregone for the duration of the festival; it may be performed using wet wipes or may take place in temporary communal or individual showers provided at larger or boutique festivals. This has a number of consequences for energy use when considered in relation to how washing is performed outside the festival. The permanent energy infrastructure and housing stock of the UK provides immediate access to hot water, and work has been done on the impact that the increasing regularity of washing in the UK has had on energy use (Shove 2004). Showering at festivals is a less frequent practice than it is in everyday life outside festivals (although it is likely recruiting increasing numbers of practitioners) and so the energy demand of washing is reduced during the period of the festival. Transformations in practices are therefore not independent of changes in accessibility to supporting infrastructures.

The relationship between permanent infrastructures, temporary infrastructures and social practices is therefore worth extended consideration. For example, the affordances of the national grid are central to the performance of many social practices and removing access to the national grid reduces opportunities for certain types of energy use that are taken for granted at home. For example, living in tents away from power points reduces opportunities for the energy use of many practices, such as watching television and blow drying hair. In most cases, festival sites return to being fields on farms or country estates for most of the year, with no permanent energy infrastructures, unlike those found in other music venues such as an arena or stadium.

Interaction between practices and permanent and temporary infrastructures can also be seen in the example of transport. Whilst the location of urban festivals provides access to public transport infrastructure, greenfield festivals are usually better served by the road network than public transport, meaning that access is most likely gained through private vehicles (e.g. cars or camper vans) or specially provided coaches from nearby urban centres or, for larger festivals (e.g. Leeds Festival or Glastonbury), by specially laid on direct coaches from more distant urban centres. The development of festivals into their current forms and the accompanying problem of audience travel, which according to recent studies is the single largest contributor to CO_2 emissions at festivals (Johnson 2015), are, in part, influenced by the relationship between festivals, the national transport infrastructure and changes in everyday transport practices (e.g. increases in car ownership and use) that form a dynamic relationship with these infrastructural changes.

The 1973 Department of the Environment report (Advisory Committee on Pop Festivals 1973) into the phenomenon of music festivals mentions trains as a primary mode of transport for festival-goers, and other sources (e.g. Anderson 1960; Sandford and Reid 1974) refer to the prevalence of the practice of hitchhiking to festivals in the 1960s and 1970s. Today, many festivals encourage a shift away from using private cars to more sustainable means such as public transport and even cycling (WOMAD 2016). Issues of audience travel also apply to urban music festivals but due to the usual presence of public transport systems and, as is often the case, lack of sufficient centralised parking, these issues can often be overlooked. That is, the permanence of established infrastructural systems provides different opportunities and constraints for urban festival organisers regarding transport compared with greenfield festivals.

As this section has illustrated, understanding the energy implications of temporary events requires attention to not only the temporality of event practices themselves, but also the (dis)continuities of their spatialities, and the temporal dynamics of their supporting infrastructures.

4.4   Conclusion

This chapter has argued that unpacking different characteristics of temporariness is an important way of distinguishing between forms of organised events. The understanding of temporariness presented here is one that moves beyond a purely temporal understanding based on duration to include such spatial aspects as the location of events and the degree of infrastructural (im)permanence exhibited by the events. Expanding temporariness in this way highlights the influence of these additional characteristics upon one another, social practices and, thereby, energy use.

The characteristics of duration, rate of recurrence, location and infrastructural arrangements interact with and affect each other so that their influence upon energy use cannot be fully understood in isolation or without reference to the historical development of events. An organised event will not rely more on makeshift infrastructural arrangements simply because it has a certain duration or rate of recurrence, nor will an event have a particular duration simply because of its location. Rather, the social practices at events are influenced by the interactions between all of the above features. For example, camping practices at greenfield festivals are influenced by the duration (more than one day) and location (usually in the countryside) of the festivals. In turn, the practice of camping incorporates parts of the makeshift infrastructure (tents) into its materiality. The need for the makeshift infrastructure that supports camping is, in part, due to the annual rate of recurrence that is part of the historical development of the events. The argument comes full circle, back to duration, when we consider proposals in the 1970s for the creation of permanent festival sites (Sandford and Reid 1974): if these proposals had been carried out, perhaps camping practices would have been dispensed with in favour of more permanent arrangements suitable for longer-term stays.

What emerges is a more complex picture of continuities and discontinuities in the practices at greenfield and urban music festivals; one that ultimately has consequences for how we think about energy use at festivals. Rather than being a straightforward case of high energy using practices taking place in a festival space, we find that many of the features of greenfield music festivals mean that the social practices that are part of the event bundle support a lower energy lifestyle than is generally the case in everyday life in our society. We also see the ambiguity surrounding our ability to clearly define the boundaries of the urban festival in space and time, creating a further ambiguity surrounding whether some of the higher energy use practices are actually part of the festival in the first place, which has consequences for how we understand the energy use of these festivals. This is not to argue that greenfield festivals are entirely low energy in their performance or that urban festivals are exceedingly high energy in theirs; transport being a major case in point. Rather, it is to argue that the consideration of temporariness and social practices allows us to see a greater complexity with regard to energy use that problematises some of the oversimplified understandings of the energy use of music festivals in the literature as well as in popular discourse.

Social practices and the temporal and spatial features of temporariness exist in dynamic relation to one another and this shapes how energy is used at organised events. By focusing on social practices and taking seriously the need to contextualise organised events within the overall picture of energy demand, it is possible to arrive at a deeper understanding of the energy demand of organised events and related issues of flexibility and variability in energy demanding practices.

Acknowledgements This work was supported by the Engineering and Physical Sciences Research Council [grant number EP/K011723/1] as part of the RCUK Energy Programme and by EDF as part of the R&D ECLEER Programme.

Bibliography

Advisory Committee on Pop Festivals, Great Britain Department of the Environment. 1973. *Pop festivals: Advisory committee on pop festivals report and code of practice.* London: H.M. Stationery Office.

Anderson, R. 1960. We were really with it. *The Observer (1901–2003).* 08/07/1960. p. 7.

Anderton, C. 2008. Commercializing the carnivalesque: The v festival and image/risk management. *Event Management* 12: 39–51.

Blue, S. 2013. *Scheduling routine: An analysis of the spatio-temporal rhythms of practice in everyday life*. PhD, Lancaster University.

Bottrill, C., G. Lye, M. Boykoff, et al. 2007. *First step: UK music industry greenhouse gas emissions for 2007*. http://www.juliesbicycle.com/files/2008First-Step-JB-report-revised09.pdf. Julie's Bicycle.

Bowdin, G.A.J., J. Allen, W. O'Toole, et al. 2011. *Events management*. 3rd ed. London: Elsevier.

Cass, N., and J. Faulconbridge. 2016. Commuting practices: New insights into modal shift from theories of social practice. *Transport Policy* 45: 1–14.

Getz, D. 2005. *Event management and event tourism*. 2nd ed. New York: Cognizant Communication.

Hinton, B. 1995. *Message to love: The isle of wight festivals, 1968–70*. Chessington: Castle Communications.

Hommels, A. 2005. Studying obduracy in the city: Toward a productive fusion between technology studies and urban studies. *Science, Technology, & Human Values* 30: 323–351.

Hui, A. 2012. Things in motion, things in practices: How mobile practice networks facilitate the travel and use of leisure objects. *Journal of Consumer Culture* 12: 195–215.

Jalas, M., and J. Rinkinen. 2016. Stacking wood and staying warm: Time, temporality and housework around domestic heating systems. *Journal of Consumer Culture* 16: 43–60.

Johnson, C. 2015. *The show must go on: Environmental impact report and vision for the UK festival industry*. http://www.powerful-thinking.org.uk/site/wp-content/uploads/TheShowMustGoOnReport18..3.16.pdf

McKay, G. 1996. *Senseless acts of beauty: Cultures of resistance since the sixties*. London: Verso.

———. 2005. *Circular breathing: The cultural politics of jazz in Britain*. Durham/London: Duke University Press.

Montagu, E. 2000. *Wheels within wheels: An unconventional life*. London: Weidenfeld and Nicolson.

Sandford, J., and R. Reid. 1974. *Tomorrow's people*. London: Jerome Publishing Company.

Schatzki, T.R. 1996. *Social practices: A wittgensteinian approach to human activity and the social*. New York: Cambridge University Press.

———. 2010. *The timespace of human activity: On performance, society, and history as indeterminate teleological events*. Lanham: Lexington Books.

Shove, E. 2004. *Comfort, cleanliness and convenience: The social organization of normality*. Oxford: Berg.

Shove, E., M. Panzar, and M. Watson. 2012. *The dynamics of social practice: Everyday life and how it changes*. London: Sage.

Shove, E., G. Walker, D. Tyfield, et al. 2014. What is energy for? Social practice and energy demand. *Theory, Culture & Society* 31: 41–58.

WOMAD. 2016. *Getting there*. Available at: http://womad.co.uk/useful-info/getting-there/

Yli-Kauhaluoma, S., M. Pantzar, and S. Toyoki. 2013. Mundane materials at work. In *Sustainable practices: Social theory and climate change*, ed. E. Shove and N. Spurling, 69–85. London: Routledge.

Michael E. P. Allen is a PhD researcher in the DEMAND Centre and Department of Sociology at Lancaster University, UK. His research looks at the energy use of live music events. He is particularly interested in how the temporal and spatial features of organised events influence social practices at events and shape energy demand.

Part 2

Unpacking Meanings

I love this garage. I'm in here all the time. It's my space, well sometimes anyway. I can mess around, make things, play some music, fix the bicycle. It's so full of stuff there's no way we could get a car in here, even if we had one. It has sort of become another room, a bit inside and a bit outside, especially in the summer when I can open up the creaky old door. Bloody cold in the winter though. It's hard to find much peace in the rest of the house, hard to make room for doing things. Everything seems to go on everywhere, all mad rush and no relaxing. In the garage time stands still a bit, feels like the hours take longer to go past, feels more like it used to. You know if you asked me what I like most about my house I would say the garage – doesn't make much sense does it!

In being interested in the dynamics of space, time and energy and their social transformations, we have to remember that ordinary (and extraordinary) lives are caught up in these dynamics. People do their best to make sense of change while they are immersed in it, caught up in the ongoing flow, sometimes embracing it and sometimes resisting. Quite a few of the chapters in this book draw on people talking about their experiences and particular aspects of their lives—sometimes the everyday routine, sometimes the more exceptional—in order to develop insights into how the demand for energy underpins and runs through the ever moving social world. People evidently can reflect upon even the most routine, habituated and mundane activities and practices that they undertake, providing an important route for researching what people do, and how and why they do what they do. Such reflection can also centre on the meanings that people find in how they live, the spaces and places they inhabit and move through and how they understand the slow, the fast, the near and the distant and other relational senses of time and space, including as they shift and evolve over time.

While the two chapters in this part of the book are not alone in being interested in how people make sense of their situated life-worlds, we have put them together because of their common concern for drawing out aspects of meaning and experience in close detail. Both chapters have the ambition of developing better analytical understandings of the shifting complexities of contemporary lives, unpacking established frameworks and assumptions in order to better capture the empirical realities they find in their data. Both are also interested in what accounts of the shifting everyday can tell us about bigger processes of social change and their differentiation, extolling those concerned with processes of intervention— to reduce the use of energy in the home, or promote more sustainable forms of transport—to better recognise the complexities and dynamics of what they are attempting to intervene into. Hanging on to well-worn, simplifying views—for example of cars as fast and bikes as slow, or homes as divided up into neat functional spaces—might have the comfort that comes with putting familiar phenomenon into familiar boxes, but isn't going to get very far when actual social experience jars and fails to fit.

For Katerina Psarikidou the challenge is to immerse a fuller set of meanings into how we understand the temporalities of contemporary mobility practices, taking on both atemporal views of car systems and their alternatives, as well as frames that work with apparently clear cut dichotomies based on simple clock-time measures. Even in ancient Greek thinking, she observes, a distinction was made between 'chronos' that sees time as measurable and quantifiable, giving centrality to the clock, and 'kairos' which sees time as experienced and qualitative in character. She argues that a kairological, polychronic approach, open to the multiplicity of temporal dimensions through which people experience mobility and give it meaning, is needed. Sitting on the bus is not simply wasted time, frustratingly long in its duration, but rather a time during which people do things—reading, listening to music, texting friends. Walking is not just about getting from A to B, but can be full of many other positive and negative meanings variously differentiated between places, day and night, and the seasons. Mobility emerges then as a shifting set of intersecting practices, materialities and temporalities, with the consequence that the meanings that people associate with moving around are more complex, relational and situated than dominant typologies suggest. Ambitions to promote modal shifts, refashion mobility infrastructures and electrify mobility technologies will always encounter the 'kairos' of mobility practitioners in some form. So being open to seeing time beyond clock time is pragmatically necessary, in addition to its appealing analytical sophistication.

Véronique Beillan and Sylvie Douzou have a more overtly spatial as well as temporal focus, concentrating on the home-space rather than moving beyond it. Their challenge is no less involved though, being concerned with the meanings that 'being at home' now carries and how these are increasingly disruptive of traditional formulations and narratives. Inhabitance of the home, they argue, is a social act and a way of seeing how broad societal changes—such as changing working practices and family structures—are refracted into people's everyday lives and their use of the socio-technical and cultural materialities of the domestic home-space. The central axis of their analysis, drawing on intensive empirical work in France, is again to take on simple dichotomies. The home is classically seen as inside rather than outside, private rather than public and clearly demarcated in these terms. Such distinctions, however, obscure the ways in which there are 'interpenetrations' at work and reflected, for example, in how people talk about the connectivity between inside and outside. This exists both physically, in terms of the uses made of doors, windows, shutters and balconies, as well as through forms of communication and digital exchange. The chapter also takes on expectations of mono-functionality in which domestic space is conceived and designed as a set of boundaries and thresholds to separate rooms with different functions and meanings. The shifting contemporary experience can be quite different with the kitchen opening up to the living room, the bedroom a place for eating, the garage a space for storage, laundry, DIY, sport and/or music, rather than merely for housing the car, and so on. Blurring boundaries in these ways and seeing the home in dynamic terms has a raft of implications for patterns of domestic energy demand—their aggregate scale, dynamics over time and profiles of access to key energy services—meaning that the units of analysis used to better capture ongoing and future change need some potentially radical re-renewing.

Both chapters therefore encourage us to see the interweaving of space, time and change through the accounts of those that experience (and indeed produce) them, with a view then to recognising the implications that follow for energy demand and its governance. Neither distils these in precise or pragmatic terms, establishing starting rather than finishing points in order to unpack dichotomous thinking and established categories of analysis. That in itself is an important contribution to how the ongoing making of energy demand should be approached in the future.

5

Towards a 'Meaning'-ful Analysis of the Temporalities of Mobility Practices: Implications for Sustainability

Katerina Psarikidou

5.1 Introduction

The significance of the temporal in shaping social life, and energy-demanding and resource-depleting living patterns (Adam 1998; Lefebvre 2004; Southerton 2009), has long been acknowledged in social research. Yet many discussions of sustainable mobility and moving away from the predominance of high emissions and energy-intensive automobility focus upon only a limited set of insights from these discussions. The social meanings attached to car systems, and their alternatives, are either largely atemporal, addressing meanings without acknowledging their temporality and change (Dennis and Urry 2009; Sheller 2004), focused on generalised comparisons between modes of transport (e.g. car as 'fast' and requiring limited waiting) or framed in terms of seemingly clear cut dichotomies (e.g. car as 'fast' vs. walking as 'slow') (e.g. Cass et al. 2004; Lyons and Urry 2005).

K. Psarikidou (✉)
Lancaster University, Lancaster, UK

© The Author(s) 2018
A. Hui et al. (eds.), *Demanding Energy*, DOI 10.1007/978-3-319-61991-0_5

This chapter argues for the value of taking a more relational and situated understanding of the temporality of mobility practices. It draws upon literature focusing on the mutual constitution of time and social practices, which has pointed our attention to the temporal organisation of contemporary ways of everyday living and underlined the centrality of social practices in not only consuming, but also producing time, as well as allocating it among different complex webs of competing practices (Shove 2009; Southerton 2013). Based on empirical research conducted in Birmingham, UK for the Liveable Cities project[1] it shows the multiplicity and relativity of meanings attributed to mobility practices: driving cars, taking trains and buses, cycling and walking. These meanings are embedded in people's everyday practices and are related to understanding competing mobility practices, and the materialities they involve. Investigating how the meanings that are linked to different mobility practices are situated within diverse temporal dynamics helps to contribute to better understandings of the complex socio-temporal organisation and ordering of mobility practices and the implications for pursuing changes to make systems of mobility more sustainable.

5.2   Temporalities, Mobilities and Sustainability

Few would disagree that time is a significant parameter influencing people's choice of and engagement with mobility practices. Yet the dominance of certain predominantly calculative or quantitative types of temporality has corresponded with the silencing or marginalisation of others. The perpetuation of the centrality of the clock in the manifestation of such quantitative understandings of time fails to acknowledge the diversity and plurality of times, but also the more qualitative characteristics and analyses that can be associated with it (Adam 2001; Geißler 2002). In his thesis for a 'new time culture', Svenstrup (2013) has talked about the co-existence of diverse time cultures across time and space and their distinct implications for the development of (un)sustainable ways of living (see also Geißler 2002). Along similar lines, Gell (1992) has talked about the diverse dimensions and perceptions of time in different cultural, geographical and temporal contexts. Drawing on such frameworks, this chapter addresses a wider spectrum of temporalities, unpacking the diversity of meanings and their role in challenging and potentially changing dominant bundles of mobility practices, such as those constructed around the car.

Multidimensional understandings of time, which go beyond the dominant clock-time monoculture, take into consideration both natural and cultural times and rhythms (Adam 1998; Svenstrup 2013; Lefebvre 2004). The temporal aspects of speed and acceleration (Rosa 2013; Virilio 1986), synchronisation, sequencing and continuity (Shove et al. 2009; Mattioli et al. 2013); frequency, repetition and duration (Heidegger 1977; Giddens 1984; Bissell 2007); disruption (Trentmann 2009) and flexibility (Cass et al. 2004; Svenstrup 2013), as well as rhythmicity and periodicity (Lefebvre 2004), as also manifested in the various natural or biological rhythms of day, seasonality, age and weather, can all contribute in different ways to analyses of how different mobility practices compare, as well as to people's understandings of what is meaningful about certain ways of travelling. As such, they underline the multiple temporal dimensions and meanings that contribute to understandings of both dominant and alternative mobility systems.

Time has been identified as central to the complex organisation of contemporary living and moving (Adam 1990; Urry 2000). Social scientists have suggested the need for a more 'time-sensitive' approach to sustainability theory and practice (Rau 2015; Geißler 2002) and underlined the significance of time, especially 'unnatural' or 'cultural' time, in the synchronisation of mobilities across space (Urry 2000). The role of clocks and watches has been central in the emergence of such patterns of time. It was the enabler of stabilisation and punctuality, calculability and precision that were necessary for fulfilling the travel needs of an increasingly mobile society for work and leisure.

After an initial ignorance of the dimension of time in Brutland's conceptualisation of Sustainable Development (WCED 1987), research has also underlined the environmental implications of the so-called 'unnatural rhythms' of the 24-hour society (e.g. a society that is not only restricted to the consumption of energy and resources during the day, but also during the night) (Rau 2015; Geißler 2002). Such approaches can date back to critiques of the industrial revolution and the centrality of the clock in the creation of a homogenised time monoculture dominated by the imperatives of speed and acceleration (see Weber 1930; Simmel 1903; Virilio 1986; Rosa 2013).

More recent analyses have highlighted the socio-environmental implications of time. Shove et al. (2009) have eloquently described the temporal allocation, sequencing, synchronisation and competition of bundles of social practices and their role in understanding contemporary time-demanding societies. In this context, time is perceived as central for the organisation of social practices, but is also crucial for the pursuit of change (Trentmann 2009; Shove et al. 2009; Southerton 2013). Time is a scarce resource that needs to be *saved* through the production of other commodities and unsustainable practices, but also a valuable commodity that is *used* for the fulfilment of energy-demanding and resource depleting lifestyles (Rau and Edmorson 2013; Reisch 2001).

Research has also underlined the significance of time in the study of mobility practices. Standardisation of time has been essential for not only measuring time, but also making new types of 'unnatural' or 'cultural' time, that could contribute to the punctual integration of mobility practices in a more stable time schedule and synchronised entanglement of flows and practices across space (Urry 2000). Emergent collective forms of travelling of rail and coach journeys have been considered as instrumental for revolutionary developments that would contribute to a 'killing of space' through a 'killing of time' (Zerubavel 1982), but also a differentiation in the 'sense of time' through a changed 'sense of space' (Wunderlich 2010). Such perceptions of time remain central in more recent analyses of the practice of travelling. Attention has been given to the practices of both mobile and immobile waiting as generators of unproductive, empty, wasted or inactive time (Watts and Urry 2008; Lyons and Urry 2005; Mackie et al. 2003; Bissell 2007; etc.). The value-laden character of time has also been central in the work of Cass et al. (2004) who have provided a socio-spatio-temporal approach to understanding a wider range of mobility practices, with a focus on the significance of the temporal aspects of speed and flexibility in the prioritisation of specific mobility patterns such as these related to the use of cars. Based on time use survey data, research has also focused on synchronisation through the analysis of a range of predominantly car-dependent social practices (Mattioli 2014; Anable et al. 2014).

Such analyses provide clues to the interconnections between temporalities, mobilities and sustainability. However, despite attempts to develop more complex and situated understandings of time, certain dimensions, such as speed and acceleration (see also Rosa 2013; Virilio 1986), remain dominant in the attribution of meaning and value to specific patterns of living and moving (Watts and Urry 2008; Lyons and Urry 2005; Cass et al. 2004). Also, most of the time, this research focuses on one dimension of time, thus paying less attention to how multiple dimensions of temporalities can exist in conjunction. This paper starts from people's experiences of mobility, bringing out how these are made meaningful in relation to different dimensions of temporality, as well as the wider intersections of temporalities, practices, materialities and spaces. It argues that since multiple temporalities are already embedded in people's understandings of everyday mobility, these should be more carefully represented and analysed in discussions of sustainability and possible mobility transitions. Such analysis goes beyond temporal dichotomies to unpack the variation and relativity of meanings related to different aspects of temporality. It thereby contributes to the development of a more 'kairological' approach (Cipriani 2013; Szerszynski 2002). In contrast to the Greek word of 'chronos' which refers to time as a measurable and quantifiable entity (e.g. year, duration, length of time), 'kairos' is associated with a more qualitative, teleological, experienced and meaningful approach to time that aims to address the properness or rightness of a time for the accomplishment of certain actions (Cipriani 2013; Dewsbury 2002). Emphasising the kairological aspects of mobility practices therefore involves recognising the heterogeneous and polysemic character of meanings related to driving, walking and cycling.

The rest of the paper draws upon research[2] conducted in the city of Birmingham. With the largest population in the UK outside London, Birmingham's historic economic dependence on car manufacturing has significantly contributed to the socio-spatial organisation of the city around the car (Cherry 1994), which despite its declining role, still constitutes the mobility system which dominates Birmingham's mobility landscape (Centro 2011). Thus, although Birmingham is currently ranked as the eighth least car-dependent city in England, car use still accounts for 42.2% of daily journeys, followed by bus with 29.2%, rail with 27%, and metro and cycling with less than 2% each (Centro 2011). Such contrasting aspects, alongside an increase in the alternatives to car use across the city, make Birmingham an interesting urban socio-spatial setting for understanding the role of 'the temporal' in mobility practices. Much evidence and debate have already focused on the significant environmental impact of car driving, which at a UK-wide level contributes to the emission of 67.4 M tonnes of carbon dioxide as opposed to 2 M for rail and 4.7 M for buses (Centro 2011). Such figures underpin the reduction of car driving as an imperative for sustainability,[3] and the car-centred mobility landscape of Birmingham provides a fruitful space for exploration.

More specifically, due to the centrality of the car in Birmingham's mobility landscape, this chapter uses the specific practice of car driving as its starting point for providing an account of different dimensions of time, their interrelationships, and how these perpetuate traditional temporal dichotomies. In the following sections, after initiating such an account, I turn to developing a more relational and situated understanding of the temporality of mobility practices that can also trouble such dichotomies. In order to do so, I pay specific attention to the multiple intersections between temporalities, practices and materialities that play a pivotal role in the attribution of specific meanings, and which are central for considering more sustainable mobility practices.

5.3   Towards a 'Meaning'-ful Temporal Analysis of Mobility Practices

Traditional Temporal Dichotomies of Mobility Practices: The Case of the Dominant Car System

As our research in Birmingham makes clear, understanding the situated complexity of the city's car-dominated mobility landscape depends upon multiple temporalities and their interrelated meanings. These meanings were made explicit at various points in the interviews. Many participants mentioned speed as central in the use of the car for their everyday mobility practices. Duration, as also manifested in conditions of waiting, has been another significant aspect that shaped everyday social practices around car driving. At other points important temporalities underpinned other topics of discussion. Safety arose as a concern, and often was related implicitly or explicitly to rhythmicity and the day/night dichotomy. Participants described decreased levels of safety associated with waiting or using public transport or cycling and walking at night.

It is clear from participants' discussions that temporal changes—for example related to seasonality and weather—affect the meanings that different forms of mobility have. Discussions around comfort emerged in relation to seasonality and weather, with participants admitting their reluctance to cycle, walk or wait in 'bad' weather conditions. As one of the interviewees described with regard to the practice of cycling:

> From the day the tunnels closed,[4] I started cycling it and I have really enjoyed myself... but...the proof in the pudding is when it starts raining more and getting colder and then we will have to see...I did cycle in one day when it rained and I got absolutely soaked all over and I wasn't prepared for that really to be honest with you. So it was quite an uncomfortable day. (Male, Birmingham City Council Representative, 2013)

Such cases of rhythmicity and seasonality highlight how sensory experiences of the temporalities of different mobility practices affect their meanings—for instance, related to conditions of cold and dark. In the case of dark, a gendered distinction in the perception of temporalities of mobility practices prevails with female travellers expressing reluctance to walk or cycle at night. Age has also been another significant parameter: a lack of competences and skills, also associated with ill health, very old or very young age, seemed to decrease people's levels of confidence to use alternatives to the car.

These sensory experiences of mobility are also important when it comes to the meanings of travelling with others. Especially in the case of younger groups, aspects of convenience and comfort associated with the carrying capacity of others—for example, travelling with babies in prams or getting kids to school—have been identified as factors influencing car-based mobility choices. As one of our female focus group participants described with regard to the use of the bus for getting her daughter to the nursery:

> On a cold and wet day, it's very cold and very wet. It would really frustrate me because I would have a pushchair or I would have my daughter in my hand and there are no seats at the bus stop or like the bus stop is broke up or you see some drunken people at the bus stop and I don't want to put my child down in her seat. (Female, Resident of High Density/High Deprivation Area, 2014)

As also identified in Mattioli's work (2014), in most descriptions of journeys related to schooling, aspects of synchronisation or continuity also prevailed, with the practice of driving kids to school constituting part of a bigger bundle of social practices that both preceded and succeeded this practice (e.g. commuting to or returning from work or going for food shopping). Such journeys of synchronised practices have also been associated with flexibility across time and space and, thus, a sense of 'freedom' and 'independence' attributed to the use of the car. Such dimensions have been prevalent in the interviewees' descriptions of the trade-offs between car driving as a personal versus a collective form of travel:

> You modify your behaviour slightly perhaps because I would probably get into work a lot earlier if I wasn't with A because he has to drop the kids off and when our first child went to child minder last year I happened to drop her off something like 9 and pick her up at 5.30. Now we never get home before 6.30, so we had to stop sharing .... (Male, Solihull City Council representative, 2013)

The above examples not only give us an idea of the multiple temporalities that are involved in the configuration of mobility systems and practices, but they also point our attention to how the meanings that arise in situated practices are more complex than temporal dichotomies can imply. For example, with regard to speed, in most research, driving is analysed as faster than cycling or walking which are perceived as 'slow travel' practices. This understanding of speed is also associated with the duration of waiting, which is shorter in the car than in other modes of transport. As also evidenced above, travelling by public transport has been associated with the perception of time as wasted or inflexible, as opposed to the more solitary practice of car driving and the high levels of flexibility and freedom associated with them. A dominant interpretation of weather, seasonality and rhythmicity also contributes to the perception of driving as a more convenient, as well as a safer mobility practice.

Such temporal meanings, however, should not be considered as absolute, or in isolation from each other, as other practices are involved in shaping them—temporal meanings are always situated and relational. The dominance of specific practices is not only related to the meanings attributed to the temporalities of these practices, but also the attribution of specific meanings to other practices that can constitute their alternatives. Such complex interconnections become pivotal for understanding how temporal dichotomies can be both reinforced and challenged in everyday experience.

Challenging Traditional Temporal Dichotomies of Mobilities: Intersections of Temporalities, Practices and Materialities

The above analysis reveals that a 'kairological' understanding of time is important for understanding the significance of values and meanings in the establishment and perpetuation of certain mobility practices with specific environmental implications. However, such temporalised analyses are also crucial for acknowledging change and variability of temporal meanings over time and, thus, for understanding and pursuing change in mobility practices and systems. The Liveable Cities research findings have been pivotal in this regard. They have contributed to realising the relative meaning of the above dimensions of time and, thus, challenging the dominant interpretations of traditional temporal dichotomies and re-thinking mobility practices beyond the dominant perceptions and meanings associated with them. As already indicated, understanding the interconnections between different temporalities can be pivotal for moving in such a direction. However, unpacking the multiple intersections between different temporalities, practices and materialities can also be pivotal in understanding the more nuanced and complex nature of such temporal dichotomies, and also perhaps changing the meanings attributed to them. In the following analysis, examples of the temporal dimensions of speed, duration and rhythmicity are selected to illustrate such multiple interconnections.

Intersections of Temporalities: Changing Perceptions of Speed

Speed is one of the temporal dimensions that research participants variously addressed. In doing so, they challenged dominant understandings of speed by unpacking the 'fast' aspects of the 'slow travel' modes, as well as the 'slow' aspects of the traditionally perceived faster mobility practices. As will be shown in the examples below, such 'distorted' perceptions of speed have actually been attributed to the role of other temporal dimensions (e.g. duration, flexibility, seasonality) that emerged as equally important in the changing perception of this particular temporal dimension.

Much discussion has revolved around duration and its role in changing perceptions of speed. Research participants described their journeys as shorter when cycling or walking. This brings into consideration the relationship between perceiving and experiencing time, with the latter playing a pivotal role in shaping the former. How time is experienced—for example as full of activity—can thus change dominant associations—for example in this case the association of driving as fast versus cycling as slow. In this context, waiting also becomes pivotal. Following similar observations in the studies of Watts and Urry (2008), Lyons and Urry (2005) and Gustafson (2012), a shift in the experience of waiting or travel time from being wasted or empty to productive and meaningful time becomes significant in challenging dominant interpretations of speed. Many participants described waiting in or for public transport as more productive than waiting related to driving: it is a time for working or catching up with work, a fact which, in turn, changes the experience of the duration of the journey and, thus, perceptions of fast versus slow movement. Such changing perceptions of speed associated with mobile work also relate to flexibility, as also manifested in the changing institutional arrangements associated with work. As one of our interviewees explained with regard to the case of rail commuting:

> It's a shorter amount of time because I am on flexi-time...I actually got more time to work...having 20 minutes downtime to actually work or read a book is probably quite valuable, isn't it? (Female, Urban Planning Company representative, 2013)

Thus, flexibility has been another temporal dimension with a significant impact changing the experience of speed and duration of time. Despite dominant associations of freedom with driving, for many participants of our research, the perception of cycling as a fast practice has significantly contributed to increasing levels of freedom associated with it. The role of digital infrastructures supporting route planning has been essential in such a transformation. However, 'faster' perceptions of cycling as well as walking have also been associated with the flexibility that non-car modes of travel can offer in the adverse conditions associated with the temporal aspects of seasonality and weather. In this case, an intersection between the temporalities of speed, flexibility and seasonality emerges. Take for example snow – while it constitutes a cause of 'disruption' for various mobile devices and immobile infrastructures that facilitate the use of car, as one of our interviewees described, it provides the opportunity for undertaking other, more flexible alternatives, and thus challenging perceptions of fast movement associated with the use of car:

> What I find very interesting about Birmingham is that it's quite a snowy City; because it's high we get snow...I love it...The way you do is it snows and you say, oh f**k and that's it. But the wonderful thing about snow is that it basically means that cars are off the road...there is nothing more pleasurable and faster than walking down the middle of Hagley Road or cycling. (Female, Local Campaigning Organisation, 2013)

Thus, the changing perception of certain temporalities—such as this of speed—is the outcome of its intersection with other temporalities—such as these of duration, flexibility and seasonality—as well as a more complex combination among such temporalities. However, as also evidenced above, such changing perceptions of temporalities is also the outcome of more complex intersections that involve different practices (e.g. working or reading a book) and materialities (mobile devices, immobile infrastructures), intersections that will be further analysed in the sections below.

Intersections of Practices: Changing Perceptions of Duration

Duration has been another temporal dimension addressed by our research participants. As becomes prevalent in the case of rail commuting, the synchronisation of practices, such as of reading or working while travelling, appeared to have significant implications in changing the meaning of different temporalities—in this particular case associated with duration. In addition to intersections of temporalities that are present in relation to duration, intersections between different practices can also play a pivotal role in shaping as well as changing perceptions of duration.

For example, focus group participants described waiting as an active and pleasant practice due to their engagement in new forms of sociality of physical or virtual co-presence with friends or strangers (see also Boden and Molotch 1994). They described it as time they could use for doing one or more things that they found valuable—for instance listening to music, playing games on the phone, reading the news and reading a book—and that would therefore add value to their practice of waiting itself and change the perception of duration associated with it. Similar intersections between practices have also been attributed to the practice of travelling, sometimes also perceived as a mobile form of waiting. Figure 5.1 constitutes a graphic representation of a journey of 'visiting family elsewhere' outlined by one of the focus group participants (Young Female, Resident of High Density/High Deprivation area 2014). As shown below, the intersection of various practices while travelling or waiting at the train station (e.g. listening to music, texting friends, walking around shops, eating or visiting restaurants) has been crucial for changing their perception of the duration of the journey as a 'meaning'-ful time. Interestingly, the prospect of synchronising different activities during

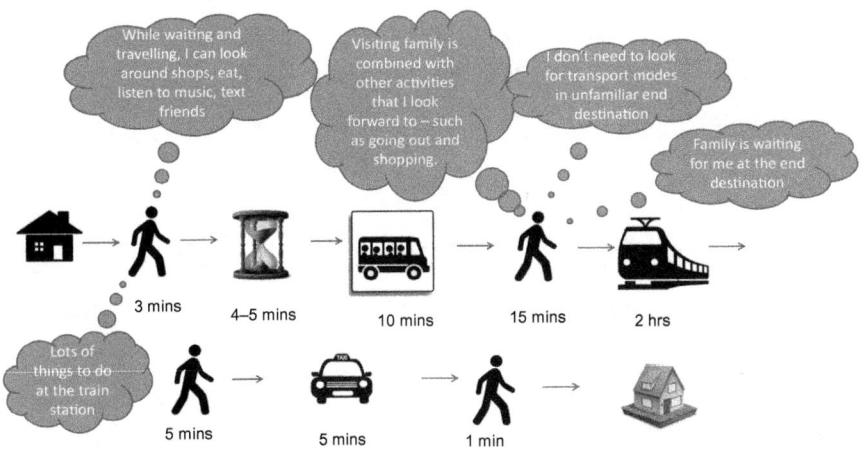

Fig. 5.1 Intersection of practices in a rail journey for visiting family

their stay at the end destination has also worked as a factor that would make the duration of the journey also more 'meaning'-ful as it could compensate for the possible 'lost time' of a non-car journey. As the focus group participant explained, the journey had not just been about visiting relatives, but also going out or going for shopping, activities that made the whole journey more worthwhile. Various materialities and infrastructures (e.g. IT devices, plugs, bookshops, cafes and other leisure spaces) have also appeared crucial for the development of such intersections between different practices and thus for transforming the experience and perception of duration.

Intersections of practices have also been central for changing the experience of duration of journeys associated with the accomplishment of other everyday mundane practices, and, thus meanings associated with duration. As already discussed above, getting kids to school constitutes a significant practice of everyday life which, as also evidenced in the 2000 British Time Use Data, has been mainly associated with driving (see also Mattioli 2014; Anable et al. 2014). However, as illustrated by one of our focus group participants (Male, Resident of High Density/High Deprivation Area, 2014), its intersection with the practice of walking the dog—another predominantly car-based social practice (Mattioli 2014)—can play a significant role

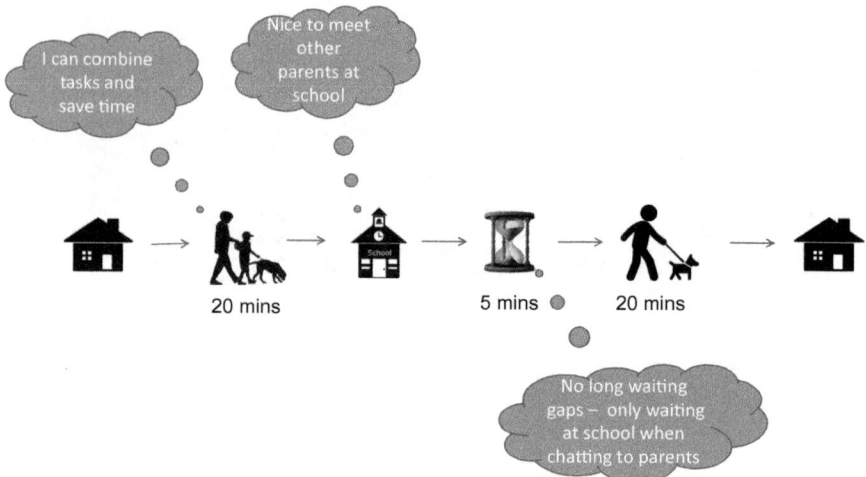

Fig. 5.2 Intersection of mundane practices of everyday living with the mobility practice of walking

in changing the perception of duration of the journey and thus the preferred mobility practice for the accomplishment of both these social practices (see Fig. 5.2).

In this context, we can also understand that walking can play a significant role in the intersection of other everyday practices, but also that such intersections of everyday practices can be pivotal for people's engagement in certain mobility practices. In this context, the experiences of 'the everyday'—as manifested in the case of schooling and walking the dog—also become important as they constitute significant contributors to changing meanings attributed to specific temporalities—such as this of duration—as well as the dichotomies related to it—for instance short versus long journey.

Intersections of Materialities: Changing Perceptions of Rhythmicity

The intersections of different materialities have also appeared crucial in shaping the experience of 'the everyday' and thus the attribution of specific meanings to dichotomies associated with other temporalities. As we will see in this section, intersections can take different forms. By intersections I do not only refer to intersections of materialities, but also of people in terms of co-existence.

Here the example of rhythmicity can provide clues to the role of such intersections in challenging dominant perceptions associated with the dichotomy of day versus night. In our focus groups, participants explained that the regular or everyday use of a specific material infrastructure—such as of a specific bus stop—and its proximity to other familiar everyday material infrastructures—such as their home or a local shop—could decrease perceptions of danger associated with the use of this materiality at night in conditions of dark. The existence of lighting infrastructures or benches also challenged perceptions of danger associated with rhythmicity and the day versus night distinction. The temporally marked meanings of infrastructure are also affected by the presence (or lack thereof) of other people. As one of our female interviewees described about her experience of Hong Kong:

> The one time in my life I've ever felt completely liberated was when I first went to Hong Kong...I felt absolutely great there...The reason was I was out and about and it didn't worry me. It is very bright there. My daughter who was then 16 flew out to join me and before she came I was quite worried because...she would have to sit in the hotel room in the evening. Not a bit; because the streets are safe and the reason why the streets are safe is because they are full of people. Streets that are full of people are safe and not just that they are safe, they also feel safe. (Female, Local Campaigning Organisation representative, 2013)

In other words, intersections between people or what Mattioli et al. (2013) would call societal synchronisation are also central for changing perceptions of safety associated with rhythmicity. Intersections between materialities are also central for facilitating such intersections between people that are, in turn, fostered through more complex intersections between people and materialities. The example of Bristol's 'Hello Lamp Posts' is indicative. As one of our interviewees described, the introduction of the 'Hello Lamp Post' in existing infrastructural public spaces contributed to changing perceptions of safety not only due to the creation of artificial day conditions, but also due to the development of a human–non-human interaction:

> Bristol ran a project called Hello Lamp Posts…you could send text messages to seats or lamp posts or bridges or other bits of street furniture and they would send you texts back…What this is allowing you to do is just interact with the City's furniture in a different sort of way. It's more of a game….The results of it seemed to be that people enjoyed texting the lamp post, texting the park bench and having these text back to them. So all of that again is an encouragement to walk around the city, explore the city and not sit in buildings and not driving cars. And all this makes the city feel safer. (Male, Multinational technology Company representative, 2013)

As evident from this discussion, an intersection between people and materialities also becomes pivotal. Mobile technologies constitute another example that has been variously used by both interview and focus group participants. Speaking on the phone or messaging friends can be seen to contribute to feelings of safety in conditions of night travelling. This intersection of specific materialities and people has been creating a digital intersection with other people and this can significantly challenge people's perceptions of rhythmicity (night as unsafe) and thus experiences of specific mobility practices that have been associated with such temporal perceptions. Of course, here, concerns over a potential digital divide, especially linked to age and the possible lack of digital competences and skills associated with older generations, can raise questions around the potential of such materialities to transform experiences of time for all.

5.4   Temporalising Mobility Practices: Implications for Sustainability

This chapter has attempted to contribute to kairological and polychronic understandings of time by focusing on the multiple interrelated temporalities of mobility practices. It demonstrated not only the multiple dimensions and interpretations of time that arise in people's accounts of being mobile, but also the complex interdependencies between different dimensions of the temporal. It has established that while there are dominant meanings of how temporal dichotomies relate to mobility practices, these are challenged by the multiplicity and relativity of meanings people attribute to their practices. To this end, it has also underlined the relational nature of temporal meanings—how different temporalities intersect with and shape each other, as well as other practices and materialities. By focusing on the examples of speed, duration and rhythmicity, the chapter has highlighted the complex relationships between temporalities, practices and materialities.

Considering the significant environmental impact of dominant, excessively car-centred, mobility systems, these discussions highlight the importance of recognising that the meanings people associate with mobility practices are often more complex than dominant typologies suggest. The analysis has suggested that sustainability efforts should not only be rooted in the transcendence of the conventional understandings and experiences of time in everyday mobility practices—for instance finding ways for cycling to be faster when compared to driving. Sustainability needs to also be explored in relation to the more complex intersections between temporalities, practices and materialities. Given the existing complexity of meanings around mobility practices and temporalities, careful thought needs to be given to how these might be changed and with what implications for sustainability.

Further research needs to develop a better understanding of such interconnections within a temporalised analysis of mobility or other types of everyday practices. However, policy and institutional arrangements need to also capitalise on such an approach, which would help us to understand mobility or other (un)sustainable practices as temporal practices and would pay more attention to the role of complex, situated understandings of time in the pursuit of sustainability. Future policies need to encourage the challenging of such traditional temporal dichotomies in order to create the conditions that can facilitate such changing perceptions of time and mobility practices. This constitutes a major challenge that needs to be addressed.

Acknowledgements This chapter draws on research conducted for the Engineering and Physical Sciences Research Council (EPSRC) Liveable Cities Programme (grant agreement EP/J017698/1). Thanks are extended to the interview and focus group participants of this research for their time and very valuable input. The author is grateful to Professor John Urry who was co-investigator in the research programme and her research supervisor. Many thanks are also extended to the two anonymous reviewers and the editors of the book, especially Dr Allison Hui for her insightful feedback.

Notes

1. See http://liveablecities.org.uk/
2. Data comes from a series of 28 interviews with representatives of organisations that may have a central role in shaping and transforming Birmingham's mobility landscape ('mobility innovators') and a series of 8 focus groups with residents of two different wards in the city of Birmingham with contrasting levels of density and deprivation. This research has been conducted as part of the EPSRC Liveable Cities research grant – agreement number EP/J017698/1.
3. For further analysis of the role of the car in current mobility systems, you may also wish to see the chapter by Mullen and Marsden in this collection.
4. The A38 St Chad's and Queensway Road tunnels that go through Birmingham City Centre were closed over the summer of 2013 for refurbishment.

Bibliography

Adam, B. 1990. *Time and social theory*. Cambridge: Polity Press.

———. 1998. *Timescapes of modernity: The environment and invisible hazards*. London: Routledge.

———. 2001. The multiplicity of times: Contributions from the tutzing time ecology project: Introduction. *Time & Society* 10: 349–350.

Anable, J., G. Mattioli, and K. Vrotsou. 2014. The mobility intensity of everyday practice: Identifying sequences of activities in terms of their travel characteristics. *DEMAND Centre Presentation*, Lancaster, 29. Available at: http://www.demand.ac.uk/wp-content/uploads/2014/09/IATUR2014_Anable_Mattioli_Vrotsou-Mobility.pdf.

Bissell, D. 2007. Animating suspension: Waiting for mobilities. *Mobilities* 2: 277–298.

Boden, D., and H. Molotch. 1994. The compulsion to proximity. In *Nowhere: Space, time and modernity*, ed. R. Friedland and D. Boden. Berkeley: University of California Press.

Cass, N., E. Shove, and J. Urry. 2004. Transport infrastructures: A social-spatial-temporal model. In *Sustainable consumption: The implications of changing infrastructures of provision*, ed. D. Southerton, H. Chappells, and B. van Vliet, 113–129. Cheltenham: Edward Elgar.

Centro. 2011. *Annual statistical report 2011*. Birmingham: Centro.

Cherry, G.E. 1994. *Birmingham: A study in geography, history, and planning*. Chichester: J. Wiley.

Cipriani, R. 2013. The many faces of social time: A sociological approach. *Time & Society* 22: 5–30.

Dennis, K.L., and J. Urry. 2009. *After the car*. Cambridge: Polity Press.

Dewsbury, J. 2002. Embodying time, imagined and sensed. *Time & Society* 11: 147–154.

Geißler, K.A. 2002. A culture of temporal diversity. *Time & Society* 11: 131–140.

Gell, A. 1992. *The anthropology of time: Cultural constructions of temporal maps and images*. Oxford: Berg.

Giddens, A. 1984. *The constitution of society*. Cambridge: Polity Press.

Gustafson, P. 2012. Travel time and working time: What business travellers do when they travel, and why. *Time & Society* 21: 203–222.

Heidegger, M. 1977. Building dwelling thinking. In *Basic writings: From being and time (1927) to the task of thinking (1964)*, 1st ed. New York: Harper & Row.

Lefebvre, H. 2004. *Rhythmanalysis: Space, time, and everyday life*. New York: Continuum.

Lyons, G., and J. Urry. 2005. Travel time use in the information age. *Transportation Research Part A: Policy and Practice* 39: 257–276.

Mackie, P., M. Wardman, A. Fowkes, et al. 2003. *Value of travel tome savings in the UK*. Leeds: Institute of Transport Studies. Working Paper 567. Available at: http://eprints.whiterose.ac.uk/2079/2/Value_of_travel_time_savings_in_the_UK_protected.pdf

Mattioli, G. 2014. Car dependent practices: Initial findings from a sequence pattern mining study of 2000 British time use surveys. *Writing Transport Geography*, ITS Leeds, 2nd April 2014, 22. Available at: https://tgrg.files.wordpress.com/2014/05/mattioli_slides.pdf

Mattioli, G., E. Shove, and J. Torriti. 2013. *The temporal and societal synchronisation of energy demand.* Available at: http://www.demand.ac.uk/wp-content/uploads/2014/02/The-timing-and-societal-synchronisation-of-energy-demand-for-web-1.pdf

Rau, H. 2015. Time use and resource consumption. In *International encyclopedia of the social and behavioural sciences,* ed. M. Fischer-Kowalski, H. Rau, and K. Zimmerer. Oxford: Elsevier.

Rau, H., and R. Edmorson. 2013. Time and sustainability. In *Methods of sustainability research in the social sciences,* ed. F. Fahy and H. Rau, 173–190. London: Sage.

Reisch, L.A. 2001. Time and wealth. *Time & Society* 10: 367–385.

Rosa, H. 2013. *Social acceleration: A new theory of modernity.* New York: Columbia University Press.

Sheller, M. 2004. Automotive emotions feeling the car. *Theory, Culture & Society* 21: 221–242.

Shove, E. 2009. Beyond the ABC: Climate change policy and theories of social change. *Environment & Planning A* 42: 1273–1285.

Shove, E., F. Trentmann, and R.R. Wilk. 2009. *Time, consumption and everyday life: Practice, materiality and culture.* Oxford: Berg.

Simmel, G. 1903. The metropolis and mental life. In *The city cultures reader,* ed. M. Miles, T. Hall, and I. Borden, 2nd ed., 12–19. London: Routledge.

Southerton, D. 2009. Re-ordering temporal rhythms: Co-ordinating everyday practice in the UK in 1937 and 2000. In *Time, consumption and everyday life: Practice, materiality and culture,* ed. E. Shove, F. Trentmann, and R.R. Wilk, 49–66. Oxford: Berg.

———. 2013. Habits, routines and temporalities of consumption: From individual behaviours to the reproduction of everyday practices. *Time & Society* 22: 335–355.

Svenstrup, M. 2013. *Towards a new time culture – conceptual and perceptual tools.* Available at: https://time-culture.net/wp-content/uploads/2013/03/Towards-a-new-time-culture-A4.pdf

Szerszynski, B. 2002. Wild times and domesticated times: The temporalities of environmental lifestyles and politics. *Landscape and Urban Planning* 61: 181–191.

Trentmann, F. 2009. Disruption is normal: Blackouts, breakdowns and the elasticity of everyday life. In *Time, consumption and everyday life: Practice, materiality and culture,* ed. E. Shove, F. Trentmann, and R.R. Wilk. Oxford: Berg.

Urry, J. 2000. *Sociology beyond societies: Mobilities for the twenty-first century.* London: Routledge.

Virilio, P. 1986. *Speed and politics: An essay on dromology*. New York: Semiotext(e).

Watts, L., and J. Urry. 2008. Moving methods, travelling times. *Environment and Planning D: Society and Space* 26: 860–874.

WCED. 1987. *Our common future*. New York: Oxford University Press.

Weber, M. 1930. *The protestant ethic and the spirit of capitalism*. London: George Allen & Unwin.

Wunderlich, F. 2010. The aesthetics of place-temporality in everyday urban space: The case of Fitzroy square. In *Geographies of rhythm: Nature, place, mobilities and bodies*, ed. T. Edensor, 45–57. Farnham: Ashgate.

Zerubavel, E. 1982. The standardization of time: A sociohistorical perspective. *American Journal of Sociology* 88: 1–23.

Katerina Psarikidou is a Knowledge Exchange and Senior Research Fellow at the Sociology Department and Centre for the Study of Environmental Change at Lancaster University, UK. Her research revolves around a multifaceted exploration of the innovation and sustainability potential of alternative agro-food and mobility practices. She is currently working for the N8 AgriFood Resilience research programme funded by the Higher Education Funding Council for England (HEFCE). She has experience in interdisciplinary, cooperative and knowledge exchange research programmes (e.g. European Commission Framework Programme 7 'Facilitating Alternative Agro-Food Networks: Stakehoders Perspectives on Research Needs' (EC FP7 FAAN), European Commission Framework Programme 6 'Participatory Governance and Institutional Innovation' (EC FP6 PAGANINI), Engineering and Physical Sciences Research Council (EPSRC) Liveable Cities). Her research has been published in journals such as *Ephemera, Sustainability: Science, Practice and Policy, Sustainability,* and *International Journal for the Sociology of Agriculture and Food*.

6

Being at Home Today: Inhabitance Practices and the Transformation and Blurring of French Domestic Living Spaces

Véronique Beillan and Sylvie Douzou

6.1 Introduction

The domestic space is a space of significant investment both in terms of time spent at home and the size of budget devoted to home-living. In most European countries, housing is the main budget item, well ahead of food,[1] with variations according to locality and the socio-economic class of the occupiers. Yet it is a place very much invested with symbolism, a preferred place for the transmission and (re)production of socio-spatial and socio-temporal norms. So whilst a habitat is a physical space, an infrastructure, an object and a well-defined and fully objectivised fact, *inhabitance* is a social act that is often overshadowed by the habitat and its economic and functional dimensions (Lefebvre 1974). Hence what "being *at* home" means today is for several reasons a *social marker* of how current societal changes become incorporated into people's daily lives and, consequently, we argue in this chapter, a way of deciphering shifting

V. Beillan (✉) • S. Douzou
EDF R&D, Palaiseau, France

patterns of energy demand and their underpinning logics, rationalities and related dynamics.

The home is first and foremost a shared social living space, marked by social practices, habits, multiple rhythms and temporalities, as well as a *locus* incorporating social dynamics and people's aspirations in terms of models and life plans (see in particular Durand-Daubin 2016a, b; Filiod 2004; Staszak 2001; Wilhite et al. 2001). It is also, in itself, a complex socio-technical system where material culture, artefacts and various technical objects intermingle. Furthermore, home is a place that touches the private sphere of life, and as such is governed by its own rules and logics, but it is also a space deeply anchored in a given local territory (and its infrastructures, sociability potential and services). If one adopts such a perspective (as we are doing), space (including the domestic) is not reduced to being a mere mirror of pre-existing and fixed social relationships, it is also an actor and producer of such relationships; space and the human activities that run through it co-construct one another in a dialectic process that is both complex and continually renewed (see in particular Guy and Shove 2000; Shove 2003; Shove et al. 2012).

In this chapter, we analyse and illustrate how and why various domestic spaces are caught up in some of the most significant societal transformations we can observe in France[2] (as in most Western European countries). Domestic space and the way it is organised is a support for the concrete acts, practices and gestures performed in a given space, time and context (Filiod 2004; Shove et al. 2012; Schatzki 2010). That is why Filiod (2004) prefers the notion of "universe", which may take better account of dynamics than the notion of a domestic space. The term "universe" makes it possible to see the extent to which spaces, in their material, physical and architectural definition, are grounded in what people do with them. The word "universe" allows us to affirm the role that inhabitants play during the act of inhabiting. The domestic space is thus a *universe* lived in, inhabited and dreamt by individuals and a space traversed by societal evolutions which also help to shape it, particularly in terms of greater modularity and permeability. Changes in the working world (working hours and flexibility, working from home) and modifications to family structures due to increasing numbers of separations and blended families, are leading to greater variation in the spatial and material make-up of home settings. Such transformations have deeply modified the geography of how French domestic spaces are used and the dynamics of occupation and inhabitance practices. These changes will then impact on end-use energy demand, potentially, although not necessarily, leading to increased energy consumption.

We explore such dynamics through a study conducted among 59 households located in the Greater Paris region.[3] This research aimed at better understanding the current logics and rationales underpinning energy demand at home, recognising that energy is not used as an end in itself but in order to perform activities and access services in a given context (Shove et al. 2014). To highlight the transformations of these domestic spaces, we re-analysed a corpus of in-depth qualitative interviews of people living in the Greater Paris region in 2010 and 2011.[4] We chose to embed our analysis within a spatio-temporal approach and to use the material culture of the home (its infrastructures, appliances and artefacts) as our entry point for analysis. In what follows we focus on two themes which emerged clearly from our analysis: firstly inhabitance in context and the permeability of spaces and secondly the multifunctionality of domestic spaces and the role of temporalities in these. We conclude by emphasising the need to revise more conventional units of analysis and understandings of the domestic space in order to better decipher and take into account current energy demand dynamics.

6.2 Inhabitance in Context and the Permeability of Spaces

First and foremost, inhabitance means having shelter, as can be seen from numerous metaphors for the home such as nest, bubble or cocoon (Pezeu-Massabuau 1983; Filiod 2004). One notion that is central to the anthropology of space, habitat and inhabitance is that of *limit*. Other social sciences disciplines also use this notion but refer rather to the idea of boundary or enclosure; yet they all agree on the fact that such demarcation can be just as much symbolic or social as material, because as a geographer (Staszak 2001: 345) points out, "the domestic space is above all a private space, that of the at-home". It constitutes what he calls a *fundamental territory*. The ever-present idea of limit goes hand in hand with a set of couplets which interweave in accordance with complex and changing processes of "demarcation"—inside-outside, public-private, day-night, order-disorder, clean-dirty.

The blurring between what relates to private and public spheres, the interpenetration of open and closed spaces, of inside and outside, is a central axis of our analysis. If the "at-home" is still often presented as a refuge from the outside world, and, as such, is part of a withdrawal, it can in no way be reduced to the simple function of a protective envelope against the "threats" of the outside world. Domestic space is in fact linked to numerous outside universes in many different ways; in particular, the uses made of the communication tools found in these universes, and of the services to which they provide access, are emblematic of the idea that inhabitance is in constant interaction with the outside world, in a movement which is also an "opening-out". So what we in fact see is more of an interpenetration of the outside and inside.

As the following interchange between a retired couple demonstrates, direct access to the outside is preferred, appreciated and organised by residents in relation to certain specific and almost ritualised moments:

Husband:	And the fact that there's no stairwell! When you live in a flat with no stairwell, it's a bonus!
Interviewer:	In what way?
Husband:	Well, with a stairwell you've got the smells and the noise.
Wife:	And then there's the pleasure of being able to see people to the door, you open the door and you're outside. Accompanying people to the door is very nice! It's pretty much like having a small town house. Well not exactly, but there's the nice fact of not having to … um … I think here it's like being in a building without being in a building! (Retired couple, living in a flat)

Where there is no direct access to the outside, it is something people look for and sometimes even dream about:

I really don't want to end my days here. Even though I'm not so badly off ... seeing some greenery and being able to open the door without worrying about neighbours coming in. Need space and greenery. [...] So we made a vegetable garden and we grow courgettes, tomatoes ... herbs, aubergines, radishes (Woman living in a flat, in a couple)

Doors, windows shutters, curtains, bay windows and even intercoms are all examples of interfaces that allow one to manage, regulate and negotiate the movement and circulation of actors—human or otherwise (animals, noise, air, light, smells)—between the outside and the inside. For example, a door contributes towards an occupier's feeling of physical and psychological safety and, as Kaufmann (1996) points out, may be used in diverse and more or less ritualised ways, depending on the circumstances. Material interfaces meet this need to feel safe and to insulate oneself from noise—or rather from certain noises in particular:

It's not properly insulated. Especially upstairs in my bedroom, there's damp around the Velux roof window. And then, um, it's very noisy around here. It's horrible, really, when they start honking their horns because they can't move ... of course, in the summer we open the windows because it's hot. But the kids are outside, shouting ... there's always noise. It's horrible, we can't get any rest ... because as you get older, noise is something that you cope with less and less. It drives me crazy. I cope with it less and less. I'd like to find a job and then be able to move. But then here, in some ways I feel safe. Because there's a door downstairs, there's an intercom, there are only six flats. ... I'm safe. But the noise is awful. I can't stand it any longer (Woman living in a flat, single, no children)

Windows as interfaces constitute an emblematic "boundary object"[5] which also afford "protection" from the outside, particularly from prying eyes. Yet these glass partitions also allow interpenetration between domestic/private and public spaces. The boundaries between private and public spaces are thus negotiated: blurred or, on the contrary, marked by the use to which the curtain is put. For many of our interviewees, it is more a question of filtering and organising than of blocking, so as to anchor their private space to a local social, spatial and temporal territory and to somehow maintain links with which they identify and in which they see themselves:

Interviewer:	How do you see your home, in terms of comfort, from all points of view – light, noise, temperature?
Woman:	Well … I think, yes I can say I think it's very comfortable. Especially since I changed the double-glazing over there (veranda) and in my kitchen. Because now I don't hear too much noise from outside, and that's good, and I can still hear a bit, because I wouldn't want to be cut off. I'm in a town, I live in a town, and it's important to hear where you live. From a thermal point of view, now, I think it's comfortable. (Woman living in a flat, single, no children)

Analysis of what interviewees want to let into or remove from their homes conveys the idea of an orchestrated movement between the domestic space and the outside which reminds one of waves, with their perpetual back-and-forth movement that changes with the tides and seasonal cycles. One opens in order to allow the air to circulate under certain specific circumstances and at clearly identified moments in time: to drive smells out of the kitchen, to hear noises that are considered to be pleasant or to allow cats and dogs to move freely in and out, amongst many others. Conversely, one closes or re-closes in order to protect oneself from air that is too cold or from sunshine that is too hot or from unpleasant noise, indeed from anything that affects well-being in the home[6]:

Generally speaking the house is open, when the weather is mild. The doors and windows of the house are often open. The air circulates a lot. Until recently the air came through a lot, there were holes in the walls, and I never filled them in because I really like it when the house is well-aired. Of course, we aired in the winter, but not so much because we could feel the air moving around but we didn't open the windows so much in the girls' bedroom. For me, it's a bit to do with my relationship with the environment, I need to feel that the air is circulating. I can't live any other way. (Man living in a house, in a couple, two children)

What well-being at home means is constantly negotiated, arbitrated and sometimes debated between the various actors sharing the same domestic space, re-affirming that comfort (and especially thermal experiences of it) is a social construction:

> He's always too hot in any case, so it doesn't matter. He's always in a T-shirt … I'm always in pyjamas and dressing gown and he's always opening the window in the kitchen. You were asking me which window was most often open, it's the one in the kitchen, we don't open the others, we smoke … he smokes, so he's always opening the window to smoke outside. (Woman living a house, in a couple, two children)

Doors also mark boundaries within domestic space, boundaries that are also constantly renegotiated, especially in relation to the possibility of creating spaces which can shift from the status of shared space to that of private space:

> And so when we had to … In the end, this post, that you can see here, this pretty post, it … You used to be able to walk all the way around it, you know … There was no partition, on the side, so we added a partition and I made a sort of door which closes, so that in the evening we've got our own bedroom. Because otherwise it was really very open. (Women living in a flat, in a couple, three children)

The use and positioning of doors relates to the process through which housing architecture became more 'functional' due to the development of the nuclear family (the father-mother-child(ren) model that dominated Europe from the nineteenth century through to the 1970s) (Perrinjaquet et al. 1986). Eleb and Châtelet (1997) point out that until the seventeenth century, family members shared the same space; it was the "common room" that predominated, a space where people did everything from eating to sleeping. With the possible exception of alcoves, personal privacy was virtually non-existent. It was only towards the eighteenth and nineteenth centuries that among the bourgeoisie a new relationship with intimacy began to appear in the domestic space, with separate rooms, each with its own specific purpose. The domestic space was transformed, becoming a set of rooms linked together by corridors, an architectural model of space which was then disseminated into all layers of the population. The urbanisation and technical logic which sprang from the industrialisation of societies promoted architectural functionalism, with each room having a single—or even unique—function; pushed to the extreme, this turns a home into a veritable *machine for living*, to use Le Corbusier's expression to characterise his villa Savoye built in 1929 near Paris.

The domestic norm quickly became mono-functionality, and in this mono-functional model, boundaries are clearly defined, taking the form of thresholds, with doors acting as powerful markers of the line between public and private, or order and disorder. Some rooms such as the living room are public and used to receive people; other ones such as the kitchen remain "technical" rooms behind the scenes, where disorder is hidden from the guests.

Since the 1980s, the transformation of domestic spaces has been essentially characterised by opening the kitchen onto the living room. This opening up is reflective of how families operate; with family life apparently adapting poorly to mono-functionality the living kitchen, for instance, is supposed to be better adapted to daily practices of families, giving mothers more opportunities to combine household tasks and supervision of children (Van der Schoor 2016). We observe the same kind of evolution among our interviewees:

> There's a living room joined onto the kitchen. It's not an open-plan kitchen, but the two rooms open onto one another, so it's without any doubt the main room, with an office on the floor above that is not closed off, which is a walk-through area and which would also be the second main room, especially for me and the kids. So yes, there is clearly a main room. It's the room in which we leave the lights on pretty much all the time when we're not in bed, because we're always there at one time or another. (Man living in a house, in a couple, two children)

Multi-functionality as a trend is also incorporated into and supported by a set of *artefacts* as demonstrated through the *multiple uses* of some ordinary objects. The dining table appears emblematic of such multi-functionality. It now serves many purposes in addition to eating: educational activities, DIY, writing letters and, increasingly, work. Hybrid spaces are thus developing within homes, jointly inhabited by a wide range of objects and activities relating to different spheres.

As a further important feature, the permeable nature of space can be clearly seen in how the interviewees make use of a set of intermediate spaces. From an architectural and spatial imprint standpoint, even when they have no garden, homes open up to the outside in a variety of ways: balconies, terraces, loggias and verandas. These spaces are precious to our interviewees, because they allow them to enjoy the pleasures of "being outside" (sunshine, light, fresh air, greenery) whilst at the same time maintaining control of their private space. From this point of view, the veranda is an interesting example. Once it has been fitted out – which they always are, and to a considerable extent (particularly with plants, airer, supplementary heating appliance and electric sockets), a great many daily activities take place there throughout the year. As one interviewee declared: "So this is a wood-burning stove, so that when it's not too cold … well … when I am sewing, I switch on the stove so that I can use the veranda, because I can see nice and clearly in there and it's very nice." (Woman living in a house, single, three grown children).

The garage is another interesting example, because it is more than a mere airlock between the inside and the outside, an inert space used to park a car. A garage is frequently a space for storage, a laundry area, a DIY workshop, a room for sport or music, a second kitchen:

Interviewer:	So do you have electrical appliances in the garage?
Woman:	The washing machine … and another oven … this one has electric hotplates. It's useful when we've run out of gas. Well, it was the former owner who left it here for us. He had a kitchen, a garden kitchen they call it. So there was a sink, an oven and also a small box freezer, but we don't use it very much. (Woman living in a house, part-time cohabitation with her partner, no children)

In fact, the domestic universe is made up of a diverse set of objects which contribute to a modification of the relationships between domestic and outside spaces, to their articulation, to the negotiation of their changing interfaces and to a blurring of their contours – such as through television or computer screens constantly subjecting the domestic universe to images from the outside world. The presence of such objects has always existed in one form or another according to time period. Paintings for instance already pierced fictitious holes through the walls and out into the exterior countryside or imagined spaces (Claval 2004). Information and Communication Technologies (ICTs) for domestic uses are in direct line with such practices, which they contribute to transforming. Not only is it possible, whenever one so wishes, at any time of the day or night, to invite the outside world into one's home (replay, streaming, downloads, etc.), but the arrival of social networks and their use especially among young people allows people to put their private lives and intimate moments (at least, those that one chooses to display) on view for all the world to see. We are thus in a situation where temporalities and the lines between private, public and professional space are totally blurred.

Each of the lines of analysis we have discussed up to this point evidently have implications for how energy demand is enrolled into the universe(s) of domestic living. How windows and curtains are used, where doors are positioned, how spaces are now multi-functional and full of devices that enable that multi-functionality are each in their own way significant for the flows of energy—gas, electricity, heat—that run through the domestic space and the practices that are performed within it. It is therefore clear that as being at home has evolved, so has the energy demand that is an important enabling ingredient of inhabitance.

We will in the next section examine more fully the multi-functionality of intra-domestic spaces in terms of how this relates to the shifting temporalities of home living, and in turn to patterns of domestic energy demand.

6.3   Temporalities and the Multi-Functionality of Domestic Spaces

Our analysis so far underlines the importance, particularly in relation to energy-demand dynamics, of two movements that interweave: on the one hand a renewed multi-functionality in flexible domestic spaces and, on the other hand, the individuation of activities within these spaces. These movements go hand in hand with a new composition of individual and collective daily temporalities (on this point see in particular Walker (2014)). Multi-functionality and the change in status of rooms – or of specific places within a given room—are traversed by multiple temporalities. As can be seen in the comments of this male interviewee, separated and a single parent with shared custody of two children, these might be short periods of time when third parties are invited: friends, family and children of blended families:

Interviewer:	What room do you spend the most time in?
Man:	I think it's the kitchen. As you saw, it's the main room, so more often than not we're in the kitchen. And the children tend to be in the living room.
Interviewer:	What do you call the kitchen, might that be both upstairs and downstairs?
Man:	Well, as you've seen, it's a mezzanine floor, we can chat upstairs and downstairs, it's nice. So it's between the two. We're very rarely in this room [dining room table, around which the interview took place]. I come in here if I want a bit of quiet. And at the weekend when we're all together, we eat in here. Before that, when my girlfriend lived with me, we'd all eat in here, with no TV, just some soft background music. We all chatted away and it was perfect. Because when there's a telly, it kills the conversation! (Man living in a flat, two children and a part-time child, blended family)

Temporalities can also clearly extend over longer periods of time, to the life cycle of individuals marked, for example, by their children becoming teenagers or leaving home, thus leading to reconfigurations of rooms. In the extract below this leads to a room becoming a space for nothing and therefore also for everything:

Man:	She took over her sister's room, which is normal.
Interviewer:	So what happened to this room?
Woman:	It's been turned into a sports room, there's the bicycle, there's a computer. It's used for everything and nothing!

| Man: | It's used for a bit of everything! We used it as a greenhouse this winter! We use it to hang the washing. It's a bit of a mess at the moment, because I did some clearing out in the bookcase. |
| Woman: | So there you go, it's also used as a storage area, for everything and nothing. (Retired couple living in a house, three children) |

We also see spaces becoming elastic and porous, depending on the time of life or on social time: week/weekend, seasons, and an elasticity which transforms a space by giving it a new room status (the bedroom of a child who has left home is turned into an office) or which transforms it into an ephemeral third-party space:

> Over there, that's a garage, a studio and a workshop. What I mean is that where there's the car, it's a garage, there's a little canopy, on the right, there's a little workshop where we put the bikes and the sun loungers in the winter, and up above there's the "pad" which has been renovated and insulated … there's a small electric radiator, there's something to provide electricity for the bass guitars, the amplifiers and whatever else you need … so that when they have … well, they don't call them parties any more, but rave-ups, they go over there, I take the car out, they close off the entrance with a plastic sheet to keep warm, it's often summertime, and then where the car usually goes they set up a big table and then, well, they are pretty much free to do what they want. Well, all that's missing are toilets, but that's not possible, I can't install all that. They cook a huge pan of pasta, then they all go over there and they do what they want all night…. (Woman living in a house, divorced, two children)

It should be noted that a room's change in status—from cold room to warm room, from closed room to a room that opens onto the outside, changes in accordance with intra-day (e.g., bathroom), weekly (day of the week/weekend) and seasonal temporalities: this is the case of the veranda which is equipped with an electric heater in the winter, in order to enact certain activities whilst making the most of the daylight, or of the living room, which might open onto the veranda or the garden:

Interviewer:	Which room do you spend the most time in over the course of the day?
Woman:	Well, I'd say it depends on the weather. At the moment it's the kitchen, not so much the living room because I don't watch the telly much and when it's warm on the veranda and in the summer, the two doors are always left open as soon as the nice weather starts ... because the veranda can be closed, it's got anti-intrusion windows, and we eat there. That gives us a big living room. We walk through it ... yeah, the doors stay open. And even when I have people round, to avoid the doors ... to avoid any problems, I take them off. (Woman living in a flat, single, three grown children)

Different "sign objects" thus mark the temporality of a room's change in status: electric heating and the tea trolley, for example, or the "openings" (doors, windows) which, as stated above, are not only technical temperature management devices, but also devices for managing privacy and social relations.

We cannot mention the increasing role of nomadic equipment without coming back to what is happening with the generalisation of mobile ICTs (computers and telephones in particular) and of their uses. As De Singly (2000) states, ICTs affirm the reign of being *"free together"*, in as much as the "roaming" that they are designed to serve gives new plasticity to social practices that take place in the home, starting with the work/non-work relationship. The transformations currently taking place in the French work market are characterised by the development of various forms of flexibility and by an increasing precarity that is causing many people to "invite" their professional world into the home, thus blurring boundaries which until recently had been relatively clear:

Well at the moment it's not very significant, because ... in fact I'm a salesman and it hasn't been very long ... I started 6 months ago ... my long-term objective is to work from home, to go to my appointments and to go into the office twice a month. That's all. The idea being to spend more time at home than at the office. At the moment it's a little bit the opposite, because

I'm in training, so I force myself to go to the office to instil some work discipline, etc. … which I will apply when I'm home. (Man living in a flat, in a couple, two children)

There are numerous examples in our data of the blurring of private and professional spheres, whether it is a case of an established professional activity, or work that one brings home and transmits via a more or less safe internet connection—not to mention telephone calls and text messages which can be exchanged with work colleagues at any time of the day or night. These evolutions are what Rosa and Renault (2010) consider to be the main characteristic of globalisation within so-called modern urban lifestyles; an acceleration or densification of social times, a phenomenon that goes hand in hand with powerful temporal tensions and a feeling of "shortage" of time, or even of "harassment". As he points out, technological acceleration gives us the impression that we are "saving time", but such time is used for other tasks, particularly for professional work done at home, thus establishing "urgency" as a universal social fact. The effects created by new ICTs in all spaces of modern life can thus be found in all aspects of everyday living (work, social and family relations, leisure). The consequences for energy demand are becoming readily apparent. IT/media now represent more than 26% of electricity used for lighting/appliances in French homes and it is by far the fastest growing source of energy demand (CEREN 2015). Its role in promoting the de-synchronisation of the timing of practice performances can also be read into the transformation of the morphology of the domestic energy consumption peak. For domestic users in the aggregate, synchronous uses' peaks tend to be less intense in 2010 than they were in 1986 (Durand-Daubin 2016a), suggesting less fixity and an increased flexibility in how the flow of everyday life is organised within domestic universes.

6.4   Conclusion

In this chapter, we have demonstrated that contemporary domestic spaces are far from being objectifiable and static enough for them to be captured solely through an array of physical and quantified measures. Our findings highlight that domestic spaces in France do not fit anymore into how they were dominantly represented during the twentieth century, assuming the existence of a clearly defined geographical space and the articulation of a set of distinct, fixed and well-defined rooms. This model goes hand in hand with an over-segmented understanding of daily life, compartmentalised both spatially and temporally. Instead, our findings have revealed two co-existing, entangled and ongoing movements that fundamentally challenge such assumptions: multi-functionality (versatility in the use of a single space and the blurring of conventional socio-spatial and socio-temporal boundaries) and individuation (individuals preserving their own personal spaces while maintaining some selective collective times at home). These movements result in increasingly de-synchronised and spatio-temporal context-free activities served by ubiquitous access to services and 'nomadic' tools (especially ICTs). They go hand in hand with an ever increasing quest for time saving and greater porosity between locations where daily practices are performed and that have hitherto been distinct.

These evolutions are bringing about new practices, new lifestyles and motivations (such as, e.g., the search for greater control over space-time) which themselves impact ways of inhabiting the domestic space with an increasing need for individuation. This relates both to greater autonomy for households (having multiple services within the home) and to increasing autonomy for individuals within the household (some of the appliances and domestic spaces are made more personal).

Whilst we have focused on revealing and interpreting these dynamics of change, it is clear that in various complex and multi-directional ways they have, and will continue to have, potentially significant implications for patterns of domestic energy demand—in terms of their aggregate scale, their temporal dynamics and their more local profiles of energy service access. This complexity and the breaking or blurring of conventional boundaries that we have examined imply that to track the dynamics of domestic energy use requires renewed units of analysis that would better capture the daily continuum of change as well as emergent future patterns.

Acknowledgements This work has been carried out within the framework of the DEMAND: Dynamics of Energy, Mobility and Demand Research Centre funded by the UK Engineering and Physical Sciences Research Council as part of the RCUK Energy Programme and by EDF as part of the R&D ECLEER Programme. The French case study was supported by the French National Research Agency. The authors also want to express their gratitude to the reviewers and in particular Allison Hui and Gordon Walker for their careful readings, relevant remarks and support.

Notes

1. French households spend 22 % of their gross income on housing running costs, twice as much as they spend on food and travel (INSEE 2014).
2. Among the most significant transformations we note those linked with socio-demographic trends (ageing of the population, middle classes' pauperisation); those linked with labour market evolution (unemployment, increase of part-time work and double jobs); those linked with new modes of access to leisure or training (games played across networks, streaming carried by ICTs) and those linked with the growing complexity of life cycles (increase in the number of single-person and of one-parent families, not a linear succession of single life, life in a couple, divorce and blended family).
3. Greater Paris region is the most populated in France (almost 20 % of French population) and is constituted of extremely varied territories: very urban (Paris), industrial suburbs and quite rural areas.
4. These were conducted using a semi-structured interview guide which explored people's day-to-day relationships with energy, examining their energy sufficiency in general and energy-saving practices in particular. Primary source was the ENERGYHAB project—"Energy Consumption, from home to city—social, technical and economic aspects". This four-year project was conducted in partnership with the LAVUE (Architecture, City, Urbanism and Environment Lab) and the CSTB (Scientific and Technical Centre of Building). It was funded by the French National Research Agency (ANR).
5. Star and Griesemer (1989) introduced the notion of "boundary object" in an ethnographic study of the coordination mechanism of scientific work. Since then, this notion has been greatly enhanced and used in various

research fields in order to qualify conceptual or material spaces at the interface of different social words; Trompette and Vinck (2009).

6. It is interesting to note that certain pieces of equipment which are specifically designed to control such circulation in order to increase the energy efficiency of homes, and were to a significant extent in evidence in the homes of our interviewees, are far from fully replacing (as a techno-centred stance might lead one to believe) the orchestrations of opening and closing to the outside. This is particularly the case with cooker hoods and the artificial ventilation provided by centralised mechanical ventilation systems.

Bibliography

CEREN. 2015. *Données statistiques du ceren, année 2015 – répartition par usage des consommations finales d'énergie des résidences principales (climat normal).*

Claval, P. 2004. Les ouvertures de l'espace domestique. In *Espaces domestiques: Construire, aménager, représenter*, ed. B. Collignon and J. Staszak. Paris: Bréal.

De Singly, F. 2000. *Libres ensemble: L'individualisme dans la vie commune.* Paris: Nathan.

Durand-Daubin, M. 2016a. *Activités des ménages par usage de l'énergie: Dynamiques quotidiennes et évolutions historiques.* Paris: EDF Lab Paris Saclay. (6125-1802-2016-15929-FR).

———. 2016b. Cooking in the night: Peak electricity demand and people's activity in France and great Britain. *DEMAND Centre Conference*, Lancaster, 13–15 April 2016, 13. Available at: http://www.demand.ac.uk/wp-content/uploads/2016/03/DEMAND2016_Full_paper_158-Durand-Daubin.pdf

Eleb, M., and A.-M. Châtelet. 1997. *Urbanité, sociabilité et intimité: Des logements d'aujourd'hui.* Paris: Les éditions de l'épure.

Filiod, J.P. 2004. "C'est quoi ce bazar?" Pour une anthropologie du désordre domestique. In *Espaces domestiques: Construire, aménager, représenter*, ed. B. Collignon and J.-F. Staszak. Paris: Bréal.

Guy, S., and E. Shove. 2000. *The sociology of energy, buildings and the environment: Constructing knowledge, designing practice.* London: Routledge.

INSEE. 2014. *France, portrait social.* Paris: INSEE.

Kaufmann, J.-C. 1996. Portes, verrous et clés: Les rituels de fermeture du chez-soi. *Ethnologie Française* 26: 280–288.

Lefebvre, H. 1974. *La production de l'espace.* Paris: Anthropos.

Perrinjaquet, R., M. Bassand, and P. Amphoux. 1986. *Domus 2005: Exploration prospective de l'habiter.* Lausanne: Ecole polytechnique fédérale de Lausanne, Institut de recherche sur l'environnement construit.

Pezeu-Massabuau, J. 1983. *La maison, espace social.* Paris: Presses Universitaires de France.

Rosa, H., and D. Renault. 2010. *Accélération: Une critique sociale du temps.* Paris: La Découverte.

Schatzki, T.R. 2010. *The timespace of human activity: On performance, society, and history as indeterminate teleological events.* Lanham: Lexington Books.

Shove, E. 2003. *Comfort, cleanliness and convenience: The social organization of normality.* Oxford: Berg.

Shove, E., M. Pantzar, and M. Watson. 2012. *The dynamics of social practice: Everyday life and how it changes.* London: Sage.

Shove, E., G. Walker, D. Tyfield, et al. 2014. What is energy for? Social practice and energy demand. *Theory, Culture & Society* 31: 41–58.

Star, S.L., and J.R. Griesemer. 1989. Institutional ecology, translations' and boundary objects: Amateurs and professionals in Berkeley's museum of vertebrate zoology, 1907–39. *Social Studies of Science* 19: 387–420.

Staszak, J.-F. 2001. L'espace domestique: Pour une géographie de l'intérieur. *Annales de géographie* 110: 339–363.

Trompette, P., and D. Vinck. 2009. Retour sur la notion d'objet-frontière. *Revue d'anthropologie des connaissances* 3: 5–27.

Van der Schoor, T. 2016. Energy scripts and spaces. *DEMAND Centre Conference,* Lancaster, 13–15 April 2016, 11. Available at: http://www.demand.ac.uk/wp-content/uploads/2016/03/DEMAND2016_Full_paper_128-Van-der-Schoor.pdf

Walker, G. 2014. The dynamics of energy demand: Change, rhythm and synchronicity. *Energy Research & Social Science* 1: 49–55.

Wilhite, H., H. Nakagami, T. Masuda, et al. 2001. A cross-cultural analysis of household energy-use behavior in Japan and Norway. *Consumption: Critical Concepts in the Social Sciences* 4: 159–177.

Véronique Beillan works at the Électricité de France (EDF) R&D Division, France. Having graduated in sociology from the University of Paris V-Sorbonne, Véronique develops research on housing and domestic practices, social uses of energy and innovation carried out in smart home and smart grid projects. She has been involved in the European projects Grid4EU and CityOpt on smart grids cities and in 'Socio-technical collectives and energy transition' and 'Labelling for Innovation' projects funded by the French National Research Agency. She also contributed to the research programmes 'People, Energy and Buildings' and DEMAND: Dynamics of Energy, Mobility and Demand Research Centre.

Sylvie Douzou is a Senior Social Scientist at the Électricité de France (EDF) R&D Division, France, and Core Group member of DEMAND: Dynamics of Energy, Mobility and Demand Research Centre. Educated at University of Montreal and University of Quebec in Montreal (UQUAM), she set up and led for EDF-R&D the Energy Demand and Dynamics of Consumption Programme (ECLEER: European Centre and Labs of Energy Efficiency Research) and People, Energy and Buildings Programme (UK Research Councils/ECLEER). She led the Electricité de France (EDF) social science research group on Energy, Technology and Society and collaborated on several co-funded European and French projects. Her current research focuses on the relationship between technology and social practice and the implications for energy systems and services.

Part 3

Situating Agency

Am I driving today? I don't really want to. I hate the rush hour traffic. I haven't got a meeting until 10 so I could get the bus and read the documents on the way, that'll save some time... but then again it doesn't look good if I'm the last one into work, plus if I'm waiting nearly half an hour at the bus stop like last time that's just time wasted. And then if I've got the car, I'd be able to do that client site visit this afternoon. It's much better for our working relationship to see them in person. But it's such a nice day, I really should cycle, I keep saying I need the exercise. Wait, isn't it school half term break? So the rush hour traffic won't be half as bad! Car it is, I can pick up the dry cleaning on the way home too.

The management of energy and mobility demand is an increasing concern as infrastructural capacity in many respects becomes more pressured and carbon objectives more urgent. Imperatives are increasing for overall reductions in demand, but also for spreading it, temporally and spatially, to avoid problematic peaks. Existing discourses often suggest that these objectives might be facilitated through encouraging people to make alternative choices, such as walking or cycling instead of driving, or exhorting them to do less of things such as long haul flying. Starting as we do from an interest in how social processes make energy demand, however, it is clear that such a framing of agency does little to acknowledge the interconnections between travelling and other practices. We already know that we do not have complete freedom to organise our lives as we wish:

much of what we do is caught up in strong social and institutional dynamics and in material arrangements that structure the norms and routines of everyday living. The interplay between structure and agency, however, must also be considered in relation to the interdependence of different practices, as changing how someone performs one practice changes how she can perform others. In this section, the authors therefore make contributions that stem from situating agency as not about rational choices, or even the interconnected constraints and opportunities within travel and transportation systems, but rather as enacted in relation to specific spatial relations, temporal constraints and sequences of practices. By questioning how assumptions that are routinely made about people's agency compare with participants' accounts of negotiating working, commuting, chauffeuring, caring and travelling, among other things, authors contribute to more robust understandings of the feasibility of projected or proposed changes to demand for energy-intensive mobility.

Caroline Mullen and Greg Marsden start from a concern for social inclusion and justice. Building on a literature that has emphasised that, for people in relative deprivation, transport is important for accessing important services, they suggest that there is limited scope for reducing carbon emissions among this group because of how working and commuting are interlinked. They develop this argument by highlighting the role of uncertainty in the structural arrangements of work, showing that where uncertainty over times (or spaces) for work is high, people are more likely to depend on the availability and use of a private vehicle. In the current UK context of increasingly prevalent zero hours employment contracts, as well as short-term housing contracts in a growing private rental market, this is an important point. A key conclusion is thus that forms of instability in working or living practices place significant constraints on the constitution of mobility-related energy demand and are thus particularly problematic in terms of moving away from the routine use of cars.

Julian Burkinshaw takes on moves towards increasing the flexibility of working patterns, in both spatial and temporal terms. Part of the rationale for promoting such transformations is that flexible working policies are designed to acknowledge the structures that constrain individuals' actions, and therefore increase the scope of individual agency to organise the means and timing of travel to work, and indeed whether such travel needs to be done at all. Burkinshaw's investigation identifies that what looks easy on paper does not necessarily play out this way in participants' lives, where multiple needs and workplace practices are taken into account. His engagement with concepts from practice theory leads him to examine the synchronisation and bundling of commuting practices with other household practices, revealing that journeys to work are seldom, perhaps rarely, just about journeys to work but also about school runs, exercise (or not), shopping, dog walking and being prepared in vehicular terms for activities that need to happen later in the day. The potential of flexible working policies to change patterns of commuting travel therefore needs to be understood in relation to all of these other practices with which they are connected.

Rosie Day and colleagues' challenge to assumptions is rather different in that it questions the extent to which increasing long-distance mobility and associated energy demand will actually play out as expected. Infrastructural developments and social norms that commend travel in later life are largely assumed to support a future of increasing leisure travel amongst older people. Their discussions with people moving towards and through retirement reveal that ambitions for significant travel during that time of life do exist and can be made concrete where people's circumstances allow. But, drawing upon participants' accounts, they also highlight that the agency to travel is intertwined with and limited by the form of ageing bodies and the competition for time resources posed by the needs of other people.

Something that these three chapters all expose, and discuss to varying degrees, is how energy and mobility demand are orchestrated through policies that are not generally considered to be energy policies. The timing and mode of journeys to work may be more significantly shaped by school hours than by any transport or indeed flexible work policy. Institutionally mandated flexibility, that is, uncertainty, in work hours can also strongly affect how people move around and lead to more carbon-intensive transport alternatives being used. The extent to which retired people engage in travel may be as much affected by wider social care provision—the absence of which requires them to take on more caring work—as by their own desires and financial resources. Thus, in this part of the book we can see that when seeking to either predict or manage energy and mobility demand, we need to look closely at the sets of institutional and policy structures within which energy consuming practices are embedded.

7

The Car as a Safety-Net: Narrative Accounts of the Role of Energy Intensive Transport in Conditions of Housing and Employment Uncertainty

Caroline Mullen and Greg Marsden

7.1 Introduction

Policy discourse on sustainable travel has focused on persuading people to make less use of private motor vehicles, and instead to choose shared or public transport, walking or cycling or simply travelling less as a means of reducing demand for transport energy and space intensive mobility. This discourse has been framed by transport research dominated by theories drawn from economics and social psychology in order to describe travel behaviour and to develop measures intended to encourage behaviour change (see discussion in Avineri 2012; Marsden et al. 2014; Metcalfe and Dolan 2012; Vij et al. 2013). Those theories explain travel demand and behaviour primarily in terms of individual choice (including theories of classical and behavioural economics, and social psychology).

C. Mullen (✉) • G. Marsden
Institute for Transport Studies, University of Leeds, Leeds, UK

© The Author(s) 2018
A. Hui et al. (eds.), *Demanding Energy*, DOI 10.1007/978-3-319-61991-0_7

This policy approach, and its underlying theoretical basis, is increasingly criticised on both descriptive and normative fronts from perspectives of other academic and practitioner approaches to understanding transport and travel behaviour (Marsden et al. 2014). One such challenge is found in work on social exclusion and transport, which maintains that the dominant policy approach does not sufficiently take account of how some activities and needs are especially important for people to be able to have a decent standard of living and to participate in society (recognising open questions about defining needs in this category—e.g. Mullen and Marsden 2016). This work emphasises how some forms of travel will be required to meet these needs or conduct these activities. If people are prevented or dissuaded from undertaking this travel then that should be a matter of concern to society. Further, the sort of travel required for these activities or needs is unlikely to be easily given up precisely because it is important, and so measures which do seek to change demand for this travel are likely to be ineffective (see Mattioli and Colleoni 2016) as well as potentially unjust.

This chapter develops the challenges social exclusion research presents to the dominant policy approaches to understanding and changing travel demand. In doing so the chapter provides new insights into ways in which transport policy focused on a rhetoric of individual choice may be exacerbating hardship and exclusion, and may also be counterproductive as a means of realising policy ambitions to change travel behaviour because it ignores factors which are locking in car ownership and associated high levels of energy intensive travel. An analysis of interviews with people in the north of England identifies relationships between travel needs and behaviours (in relation to often inter-related factors including journey timing, time spent travelling, ability to plan travel and decisions on travel mode) and uncertainty and insecurity in housing, employment and education. We find that uncertainty about where people live and work, and about when they need to be at work, creates conditions which encourage people to maintain or acquire a car where that is possible for them. The pressure is such that people will keep a car even where the financial resources required exacerbate households' difficulties in paying for other basic needs. For those without a car, these conditions of uncertainty can restrict opportunities and contribute to hardship. In contrast, security and a level of control over housing and employment can contribute to providing conditions in which people are more able to, and sometimes do, live without access to a private motor vehicle. We argue that the structural approach adopted in social exclusion and transport research provides a plausible and normatively defensible approach to understanding this travel demand and to assessing measures intended to alter demand. However, we make a case for a significant extension to the ways in which social exclusion research has considered travel needs. Work on transport related social exclusion is concerned with how people's choices are constrained by the systems they live in (Lucas et al. 2016). While our research recognises the relevance of structures, it extends attention from the structural context in which travel occurs, to uncertainty about how that context might change for people, and the subsequent implications for demand and changing demand.

Our findings have significant implications for transport policy intending to change travel behaviour and reduce transport related energy consumption. They add to the weight of criticism of dominant approaches which attempt to explain demand primarily in terms of choice (these are discussed in more detail below). Moreover, our findings indicate that housing and employment policy and practices are creating conditions which encourage more rather than less travel by car. As we explain in later sections, the sorts of uncertainty affecting many of our participants are becoming increasingly prevalent and their influence on transport and policy objectives are potentially large. While car ownership need not necessarily equate to increased travel, or to increased travel by motor vehicle, travel statistics suggest that there is in practice a connection. In England, households with access to a car travel longer distances (average 4580 miles/person/year) and make more trips (average 280 trips/person/year) than those without car (Department for Transport 2016: Table NTS0702). This broadly holds across income groups. While car ownership and distance travelled vary with household income (see Department for Transport 2016: NTS0703 and NTS0705), there appears to be little variation across income quintiles for the number of trips per person for each car to which a household has access. Moreover, distance travelled by each person per household car is actually higher for the lowest two income quintiles than for the third quintile, and is close to that for the fourth quintile.[1]

7.2   Choice and Context

As we have argued elsewhere, ideas of individual choice are central in three of the most prominent theories of travel behaviour and change. These theories are classical economics, behavioural economics and social psychology (Marsden et al. 2014). The idea of choice varies in each, and it is worth briefly describing these differences and the basic commonality between them. Classical economics, which is still the dominant theory of behaviour in transport policy, holds that people choose how to travel according to a self-interested, rational decision, which at any given point is based primarily on cost and time (Avineri 2012). This approach is then extended to explain individuals' broader travel decisions as influencing and being influenced by their choices about where to live and work (e.g. Kim et al. 2005; Dubernet and Axhausen 2016). Behavioural economics presents itself partly as a response to what its advocates view as classical economics' unsubstantiated assumptions about behaviour and rationality. Behavioural economics maintains that a variety of other factors influence decision-making, including loss aversion (Oliver 2012), social norms and sometimes the way in which a choice is framed (Gowdy 2008). As might be expected, approaches other than behavioural economics which also draw on social psychology treat social norms, as well as other attitudes as influencing travel choice (Shove 2010). Each of these theories recognises that context (i.e. norms or cost or time) matters in influencing choice (a point noted in Shove 2010). The theories also maintain that people are influenced by the 'nature' of their thinking, so they either think as 'self-interested' beings who follow a particular view of rationality, or they (also) think in a way which reflects attitudes. On these terms it might be argued that these theories are barely focused on choice at all and in fact leave fairly little room for choice. However, that would be to misunderstand them. In each case they focus on individuals making decisions, and then say something about the sorts of factors which might influence those decisions. If the theories are then used to try to promote change, they may suggest altering one or other of those influencing factors—such as cost, framing or social norms. It is this focus on the individual and their decision-making which characterises the sense in which choice is central to these theories.

Social exclusion and transport poverty research challenges this focus on individual choice in explaining behaviour, and consequently problematises the idea that trying to persuade people to alter their travel choices is a reasonable or effective way of bringing about change. This work tends to begin from a descriptive and a (sometimes implicit) normative concern about whether people have access to, can afford and can use the travel that they need to meet important needs and engage in important activities (Lucas et al. 2016). As we have said above, this is predicated on a position that there are some needs and activities which are required for a reasonable quality of life and level of social participation. That of course does not tell us much about what those needs or activities consist of, or how they can be met in different circumstances, and there is significant dispute on these points (for a range of approaches see Beyazit 2011; Lucas et al. 2016; Martens 2016; Mattioli 2016; Mullen and Marsden 2016; Walker et al. 2016). However, it is feasible to identify very broad categories of needs and activities with which social exclusion research is concerned. These include, but are not limited to, employment, education, housing, healthcare, family and social relations. Much of the work on transport and social exclusion to date has been concerned with availability and affordability of transport from (often low income) residential areas to services, employment, education, and so on. This body of research has brought attention to problems caused by the environments that people have to move around. For instance, it has identified harm to the health and welfare of low income women, who for reasons of transport costs, may walk in areas which are polluted and have high traffic volumes (Bostock 2001). The idea of choice may appear in work on transport poverty and exclusion, but it tends to be in the sense that choices are constrained by conditions and circumstance (Lucas 2012). Those conditions are not considered to be a matter of choice (e.g. it is not assumed that people exercise some sort of free choice to live in inaccessible neighbourhoods or to subsist on low incomes).

As we have already suggested, focusing on social exclusion potentially presents a practical and normative challenge to measures designed to change travel behaviour through economic incentive or persuasion. Yet it could be countered that those affected by transport poverty and social exclusion tend to be people with low or relatively low incomes, and perhaps measures intending to reduce use of private cars would have little relevance here if this group tend not to have access to private vehicles. However, this is not always the case: in England, latest figures show that 52% of households in the lowest income quintile have access to at least one car (Department for Transport 2016: NTS0703). Analysis of data from Germany adds to evidence that there is not a straightforward relationship between low incomes and low levels of car ownership (Mattioli 2013a, b; Mattioli and Colleoni 2016). That study suggested the likelihood of people saying they cannot afford to own a car varies according to the availability of public transport and other aspects of accessibility in the area in which respondents live. The indication is that people will maintain a car even when they struggle financially to do so, because of its importance in enabling them to get on with everyday life (Mattioli 2013b; Mattioli and Colleoni 2016). From this perspective, it can be pointless (and probably offensive) to ask people to choose conditions which enable less travel by car (e.g. to choose to live in a more accessible area), and unreasonable to ask for change in other, already restricted behaviours, such as decisions on travel mode.

7.3   The Study

We conducted 45 interviews, between 2015 and 2016, with members of the public living in West Yorkshire and Greater Manchester (in the north of England) who have gross annual household incomes up to £35,000, or up to £25,000 for a family with a single adult. The incomes approximate to the median income at the fifth decile from lowest, of households consisting of two adults and one child, and one adult and one child respectively for 2015–16 (HM Treasury 2015). West Yorkshire and Greater Manchester both comprise one major city (Leeds and Manchester), several smaller cities and numerous small towns. The interviews sought to understand not just how people travel now, but how mobility, affordability and availability of transport have affected their lives. We recruited people with a mix of car and non-car owner-ship from a range of urban, peri-urban and rural areas since we were interested in exploring how neighbourhood location and accessibility might influence travel and the impacts of transport. Participants were also chosen to have between them a mix of housing tenure types (social housing tenants; private rental tenants; owner-occupier) to allow some investigation of whether the different conditions attaching to different tenures seem to influence travel or its impacts.

The interviews were unstructured to allow a more natural discussion, although there was a topic guide and each of the interviews covered all topics. The topics then were as follows:

- How the person travels for everyday activities such as work, education, caring, leisure;
- Whether the availability of cost of travel impacts on their ability to take part in activities;
- Whether life and everyday activities are affected by road safety, or noise or air pollution from traffic;
- How transport has affected them in the past, for instance, has avail-ability of transport influenced employment, education, or housing?
- The potential implications for participants of changes in transport policy or costs—e.g. fuel price changes;
- How the costs of domestic energy affect everyday life;
- Whether housing tenure has implications for costs of domestic energy.

We do not seek here to describe the extent in wider society of the issues we discuss. What we do is identify phenomena influencing the lives of some of our participants and which are therefore potential areas for pol-icy intervention; nevertheless as we shall discuss in what follows there are reasons to think that these factors are quite widespread.

7.4   Findings: Mobility Needs in the Contexts of Housing and Employment Uncertainty and Stability

Our participants revealed strong inter-relations between travel decisions, accessibility and availability of transport, on the one hand, and uncertainty and change in housing and employment on the other. In this section we organise findings in terms of travel and employment uncertainties, and then travel and housing uncertainties. As we show in these discussions, uncertainty in employment or housing can act as a motivation to own a car, and the availability and affordability of cars can impact on welfare, quality of life and hardship across these categories. Following these discussions, we describe the situations of a few participants who find they have no need for a car.

Employment, Uncertainty and Mobility

Participants revealed some major concerns about travel and its impact on work, and on progression with work. These fall into two broad categories, the first of which is an ongoing lack of available transport to get people to what is fairly steady work in terms of location and tenure, and the second is difficulties associated with uncertainty about the location and timing of work.

For some people in work, dependence on often unreliable public transport can be a source of stress. One participant described the experience waiting for the (late) bus to work as:

> …you just start getting anxious… have I got an alternative, can I get a taxi, well how much money have I got in my purse, I haven't got any money and I can't afford a taxi. I've bought a bus pass for the month because that's what I can afford. (Helen, 40s, private renting)[2]

The difficulty of travelling without a car presents a direct barrier to progression at work. Helen also described how her commute to her work as a teaching assistant generally takes her an hour and a half each way by bus (two buses each way), and that she has to add another half hour in order to make sure that she is not late when the buses do not run to schedule. The journey would be 30 minutes by car. The length of this commute is a barrier to attending the meetings and training needed to gain promotion at the school. Moving house to be closer to work is not viable for Helen in the foreseeable future as she rents from the council and had difficulty in securing the house that she does have which is suitable for herself and her family. For another participant, travel to her former work was not a particular problem but she felt that not having her own car meant:

> I missed out on pay rises and things like that because I didn't drive ... I could sit on the phone on sales day while the others will be driving around industrial estates and stuff and you seemed to get more business if you were meeting people face to face rather than just on the phone. (Charlotte, 30s, social housing)

This was among the reasons this participant gave for learning to drive despite a personal reluctance to do so.

Travel difficulties for those in work or education are often created because location and hours change. In some cases the change occurs once or twice, when an employer or a school moves, or when hours change. This move can be, and for some participants has been, a benefit since work moved closer to home, although even in these cases there was uncertainty brought about by the prospect that the location would be moved again. However where a move occurred, it could mean that a commute which had been relatively easy on foot or by bus was now very difficult and time consuming (e.g. Thom, 60s, private renting). Further the availability, or acquisition of a car, was for two participants crucial in enabling them to manage the journey when workplaces moved (Gilly, 50s, home owner; Yasmin, 30s, private renting). For a further person, the journey after a workplace location move was also considered only possible if done by car, but was still problematic due to lack of parking (Tina, 60s, social housing—before her retirement).

For others, far from being very occasional such changes are quite frequent (e.g. Marion, 50s, social housing). This is especially, but not only, the case for people working in agencies or on zero hour contracts (these are contracts between an employer and employee where the employee has no guarantee of set hours of work, or indeed of any work). Several people explained how they have lost work because of these difficulties. For instance, one former agency worker without a car described how she would often have to turn down work offered by an agency when it was not within an area that could feasibly be reached by bus (Helen, 40s, private renting). In another case, a participant also without a car, discussed the problems he faced with a zero hours contract for which he would be sent at short notice to multiple locations and often at times when public transport was no longer running (Michael, 40s, social housing). While his employer offers to cover the cost of taxis, the expenses may often not be repaid for weeks and that causes financial problems. A further participant struggled to break into the career she sought as she had not had access to a car when she needed it. By the time she did get a car she had lost some opportunities which could have helped her begin to establish her career (Anne, 20s, living with parents). Having access to a car does not necessarily remove difficulties presented by employment hours or locations which vary due to decisions of people other than the employees. However, our study indicates that these difficulties can be greatly mitigated by access to a car or by having family members who can provide lifts. This was a reason for some participants to retain a car despite finding the costs very difficult.

Housing and Moving Home

Having been confronted with forced moves and housing uncertainty, a number of participants identified neighbourhood familiarity as important for a new home. Further, for many, fear and experience of eviction motivated a search for a new home with secure tenure. For people in this position, questions of access to and affordability of a vehicle can impact on ability to realise housing priorities, as well as on exposure to economic and other hardship.

Participants described their experiences of moving and searching for homes under a range of conditions. Some had previously owned their homes (or rather had mortgages), but have now lost those homes. In two cases, women reported that the loss of the home was because their former husbands had, without the women's knowledge, failed to pay debts on the home (Doreen, 40s, social housing; Shelagh, 50s, private renting). Another was in the process of having her home repossessed as she had been unable to pay the mortgage (Jayne, 40s). Several, who were, and in some cases still remain, in the private rented sector had faced unexpected eviction notices from their landlord.[3] Others who rent privately also move frequently, often around the same area, but appeared to think of this as something quite normal and expected at this stage of their lives (these are people in their 20s). A further participant lost a council home after the block of homes that she lived in was demolished, and while the council would rehouse her, she faced difficulties in finding a suitable replacement home from the council—her household includes relations quite distant by blood and they have found difficulties in being recognised as having housing needs as a family (Helen, 40s, private renting). Further reasons for moving included relationship breakdown and sometimes getting out of a dangerous domestic situation, bereavement, wanting to move closer to family, overcrowding, poor conditions in the home they moved from and cost.

Many people in describing decisions made when having to move home identified a priority of remaining in an area which they know and where they have established social networks. Geographical moves were also, for several people, motivated by a wish to be closer to family or friends. Experience of housing uncertainty also creates a strong influence on people's thinking about priorities. This became very apparent when people recalled their thought processes when they were in what were often very difficult and stressful circumstances of being forced to move. One of the women who reported how she lost the home she owned after her husband failed to make mortgage payments related:

I went into total meltdown. It was cost, where we would go, the area, I kind of wanted the same area but there wasn't anything. What there was, was not suitable. I'm not used to just living in anything. I wasn't accepting just anything. So that kind of went through my mind hence the fact that we're in private rental and not council. (Shelagh 50s, private renting)

The experiences of several people who were forced unexpectedly to leave a private rented house seemed to prompt them to prioritise finding a council or housing association home (both forms of social housing) which would bring security not available in the private rented sector. This is illustrated in comments by a participant who explained:

> …we'd had to move out of our last property because we privately rented and the guy who we rented it off had - I think he'd gone bankrupt or something but they gave us 2 weeks' notice to get out of the house. I think they were repossessing the house, that was it. … we didn't have any money to put a deposit down on another private rented and to be honest, I was worried about it happening again, I was like, "What if we get settled into another property-?" so yeah, we registered with the council and I think we had to live with my [extended family] for about 8 months and then we got offered this. So it wasn't by choice, it was just a better choice. (Charlotte, 30s, social housing)

These priorities appear far stronger, and more important to people, than considerations about moving to a location where work and services are accessible without a motor vehicle. That is not to say accessibility was no concern. For some having to move, accessibility of schools by public transport or on foot featured prominently in decisions. Further, several participants without cars talked about the importance of remaining close to bus routes when they moved. However this factor was raised alongside discussion of the importance to them of remaining in a geographical area to which they are attached. Further, the person quoted above also talked about how she and her family only sought council houses in locations where everyday travel would remain possible if the family had to give up the car. Again, for her and her family, this consideration sat under the priority of gaining the security of a council home following their experience of eviction from private rented housing. For some, matters of access to work are overwhelmed by family priorities. One person, without a car, moved from a place with good public transport links to his work, to a remote rural location in order to be with family, and with the result that he now frequently has to walk six miles (up and down Pennine hills) to get to work and back as there is not adequate public transport (Larry, 30s, private renting). Likewise the teaching assistant mentioned in the previous section accepted her three to three and a half hour commute to and from work by bus rather than move to a home which could not properly accommodate her family.

In summary, better access to work or other activities did not appear as the main priority in deciding where to move to once a move was forced. For those with access to a car, and for whom that car access was not at risk on grounds of affordability or illness, the priority of family over accessibility of work caused few problems. For those without a car, or who felt that their continued ownership of a car was at risk, there were some cases where the move resulted in very difficult commutes. For others, concerns about accessibility acted as a constraint on realising priorities of moving to somewhere secure, or to a familiar location.

Living Happily Without a Car

While accessibility of employment and other (non-family or friend related) activities is not generally among the priorities of people thinking about where to move to, having moved, some people found that they could live happily without a car. Some participants who had moved home and who had a car found that in their new circumstances the car was no longer needed, and became rarely used and was eventually disposed of. One such woman, with three children, moved into a new council house where she found travel to work and school relatively easy without a car. She identified that getting rid of the car does close down some activities, such as getting into the countryside; however, she added:

> But I found myself being more stressed when I had a car...I did more in terms of like the kids went to [several activities] and pretty much felt like we lived in the car quite a lot. And when I got rid of the car, they had to learn to get public transport and do that themselves. So I think I'm a lot less stressed. I read on the train and stuff. (Susan, 30s, social housing)

Two other participants also experienced a move to a place where they found the car was no longer needed in the way that it had previously been. For one family this happened when they moved to another private rented house in a new area (Hazel, 20s, private renting), and for a another participant this occurred after she bought a flat in an area she had settled in because it is close to family and feels relatively safe (Louise, 30s). For a further two participants, the decision that a car was no longer needed, and had become more a burden than a benefit, arose from a combination of retiring and remaining in an area where many services are in walking distance, and where buses provide sufficient mobility beyond that (Tina, 60s, social housing; Anna, 50s, home owner).

Among those people who moved to areas where they found they did not feel the need to have a car, there are differences which mean that some are more likely than others in the future to return to circumstances where a car might again be important to them. For one, the woman who bought a flat in what turned out to be a relatively accessible area, the issue is that her current work will soon end and she does not know what her next work will be or how feasible it will be to manage without a car. For the family who moved to a private rented house, the uncertainty is around security of tenure and where they would be able to move to if evicted from their home. This insecurity is strongly felt by the family and illustrated by their reluctance, for fear of retaliatory eviction, to complain about their landlord's recent failure to keep the boiler in good repair which led to the family suffering carbon monoxide poisoning.

For the two participants in social housing, there is no apparent uncertainty in their circumstances, and so no reason to think that they might feel they need to reverse their car free status. As a final note on living without a car, one participant, not previously mentioned here, moved into her council house 50 years ago and before retiring had worked in two places both of which were local. For her, a car was never considered or wanted, and she described her commute:

> My house was at the top of the street and I walked down and crossed…No problem… I knew everybody and everybody knew me. (Gladys, 70s, social housing)

7.5   Uncertainty, Flexibility and Prospects for Reducing Travel by Car

This study indicates the significance of understanding the conditions and circumstances in which people engage in the forms of travel that they do, and through which people can miss out on opportunities if they cannot access suitable transport. As with other work on social exclusion and transport, we have assumed that some needs and activities are more important than others, and that travel needs to respond to, and enable, these activities. Specifically we have assumed that employment, education and housing have a normative significance.

Our findings also indicate the importance security and uncertainty have in explaining travel demand and especially locking people into car ownership, and in understanding hardship related to travel. In this we have extended previous work on social exclusion and transport, showing that it is not just structural conditions, but uncertainty and insecurity about changes in those conditions which matters both in understanding what people do and where they face difficulties. The employment and housing uncertainty means that people live with the prospect of having to make complicated journeys over which they have relatively little control, and which can be the cause of substantial problems or lost opportunities. What we are seeing in part is behaviour that responds to, or anticipates, changing conditions. This includes acquiring or retaining a car in order to manage searches for work, or jobs with unpredictable locations or hours. It also involves undertaking difficult commutes which are a cost of having to moving and prioritising secure housing, or housing near family or a familiar location.

These findings identify what may be a significant challenge to policy objectives designed to reduce transport energy. As we have already indicated, the research challenges the effectiveness and fairness of policy approaches which try to alter travel behaviour by appealing to people to change their travel choices. In circumstances where people are already trying to manage uncertain housing or employment, messages about driving less are likely to be quite reasonably ignored, and by themselves, increases in costs of driving may just risk increasing hardship. Instead, uncertainty in housing and employment may tend to increase the distances people travel and the number of trips that they make because the uncertainty encourages car ownership. Conversely stability in employment and housing can open up possibilities for people to decide to give up car ownership. Our qualitative study in the north of England cannot itself show the scale of the challenge presented to transport policy by employment and housing uncertainty. However research and reports on housing and employment trends indicate that the scale could be large. People are increasingly finding themselves in the private rented sector due to high costs of buying a house and scarce social housing (Birch 2015). In 2014–15, 19 % of households in England were in private rented sector housing, compared with 11 % in 2004–5 (Department for Communities and Local Government 2016), and "76 % of private renters had lived at their current accommodation for less than five years compared to 20 % in owner occupation and 39 % in the social rented sector" (Department for Communities and Local Government 2016: 2). Coupled with this there are high levels of employee flexibility in jobs, meaning that people can be frequently looking for work, or in work that varies location and timing (MacInnes et al. 2015). A report by the Resolution Foundation found "32 per cent of the working age population (excluding full-time students) classified as *insecure* in 2014" where 'insecure' workers are those who are low paid, or part time workers, or in temporary work or have not been with an employer for sufficient time to be protected by employments rights enjoyed by other employees (Gregg and Gardiner 2015: 4). According to that report, the proportion of workers in insecure work has not greatly increased since 1994 (when it was 30%), however the authors find:

> …specific forms of atypical and often low-quality employment – including involuntary part-time and temporary working, less secure self-employment and zero hours contract working – have grown in prevalence during and since the downturn. Relatively small groups of workers (compared to the overall workforce) are affected in each case. For example, only 4 per cent of workers are involuntarily part-time, only 2 per cent are involuntarily temporary employees and only 2 per cent are on a zero hours contact (with

some overlap between these groups). But the implication is that a sizeable minority are facing particularly acute forms of insecurity. (Gregg and Gardiner 2015: 5)

Housing and employment precarity present a substantial challenge to attempts to reduce energy intensive transport and related emissions, and to tackle loss of opportunity and hardship. Unless there is a radical shift in employment and housing policy and practice, tackling this obstacle will require attention to developing a transport system in which a car is less important even where people need to make the quite complex journeys which may need to be planned at relatively short notice. This will involve developing existing ideas on how urban form, public transport and other mobility service provision can reduce reliance on motor vehicles (for instance Barbour and Deakin 2012; Mattioli and Colleoni 2016). Yet because they diverge from patterns of fairly fixed journeys at fairly fixed times, responding to travel needs in conditions of uncertainty without resort to private cars is likely to create particular challenges for planning.

Acknowledgements This work was supported by the Engineering and Physical Sciences Research Council [grant number EP/K011723/1] as part of the RCUK Energy Programme and by EDF as part of the R&D ECLEER Programme.

Notes

1. These estimates compare trips and distance for each income quintile (Department for Transport 2016: NTS0705) with the average number of cars per household for income quintiles, and assuming an average of 2.2 cars/household for households with more than one car (Department for Transport 2016: NTS0703).
2. We use pseudonyms throughout.
3. In England, many private rented sector tenancies are Assured Shorthold Tenancies (ASTs). With ASTs tenants can face eviction after six months without reason. The landlord only has to give the tenants two months' notice then gain a court order to lawfully evict the tenants. (see Shelter 2017; HM Government 1998).

Bibliography

Avineri, E. 2012. On the use and potential of behavioural economics from the perspective of transport and climate change. *Journal of Transport Geography* 24: 512–521.

Barbour, E., and E.A. Deakin. 2012. Smart growth planning for climate protection: Evaluating California's senate bill 375. *Journal of the American Planning Association* 78: 70–86.

Beyazit, E. 2011. Evaluating social justice in transport: Lessons to be learned from the capability approach. *Transport Reviews* 31: 117–134.

Birch, J. 2015. *Housing and poverty.* Joseph Rowntree Foundation. Available at: https://www.jrf.org.uk/file/47030/download?token=OCDKCeMT&filetype =full-report

Bostock, L. 2001. Pathways of disadvantage? Walking as a mode of transport among low-income mothers. *Health & Social Care in the Community* 9: 11–18.

Department for Communities and Local Government. 2016. *English housing survey 2014 to 2015: Private rented sector report.* HM Government. Available at: https://www.gov.uk/government/statistics/english-housing-survey-2014-to-2015-private-rented-sector-report

Department for Transport. 2016. *National travel survey 2015.* HM Government. Available at: https://www.gov.uk/government/statistics/national-travel-survey-2015

Dubernet, I., and K.W. Axhausen. 2016. The choice of workplace and residential location in Germany. In *16th Swiss transport research conference*, Monte Verità/Ascona, 18–20 May 2016, 16. Available at: http://www.strc.ch/conferences/2016/DubernetI_Axhausen.pdf

Gowdy, J.M. 2008. Behavioral economics and climate change policy. *Journal of Economic Behavior & Organization* 68: 632–644.

Gregg, P., and L. Gardiner. 2015. *A steady job? The UK's record on labour market security and stability since the millennium.* Resolution Foundation. Available at: http://www.resolutionfoundation.org/app/uploads/2015/07/A-steady-job.pdf

HM Government. 1998. Housing act 1998.

HM Treasury. 2015 *Impact on households: Distributional analysis to accompany budget 2015.* HM Government. Available at: https://www.gov.uk/government/uploads/system/uploads/attachment_data/file/413877/distributional_analysis_budget_2015.pdf

Kim, J.H., F. Pagliara, and J. Preston. 2005. The intention to move and residential location choice behaviour. *Urban Studies* 42: 1621–1636.

Lucas, K. 2012. Transport and social exclusion: Where are we now? *Transport Policy* 20: 105–113.

Lucas, K., G. Mattioli, E. Verlinghieri, et al. 2016. Transport poverty and its adverse social consequences. In *Proceedings of the institution of civil engineers-transport*. Thomas Telford (ICE Publishing), 353–365.

MacInnes, T., A. Tinson, C. Hughes, et al. 2015. *Monitoring poverty and social exclusion 2015*. Joseph Rowntree Foundation. Available at: https://www.jrf.org.uk/file/48631/download?token=mjrEetkk&filetype=full-report

Marsden, G., C. Mullen, I. Bache, et al. 2014. Carbon reduction and travel behaviour: Discourses, disputes and contradictions in governance. *Transport Policy* 35: 71–78.

Martens, K. 2016. *Transport justice: Designing fair transportation systems*. London: Routledge.

Mattioli, G. 2013a. Different worlds of non-motoring: Households without cars in Germany. In *Mobilitäten und immobilitäten*, ed. J. Scheiner, H. Blotevogel, S. Frank, et al., 207–216. Essen: Klartext.

———. 2013b. *Where sustainable transport and social exclusion meet. Households without cars and car dependence in Germany and Great Britain* (unpublished thesis). PhD, University of Milan-Bicocca.

———. 2016. Transport needs in a climate-constrained world. A novel framework to reconcile social and environmental sustainability in transport. *Energy Research & Social Science* 18: 118–128.

Mattioli, G., and M. Colleoni. 2016. Transport disadvantage, car dependence and urban form. *Understanding mobilities for designing contemporary cities*. Cham/Heidelberg/New York/Dordrecht/London: Springer, pp. 171–190.

Metcalfe, R., and P. Dolan. 2012. Behavioural economics and its implications for transport. *Journal of Transport Geography* 24: 503–511.

Mullen, C., and G. Marsden. 2016. Mobility justice in low carbon energy transitions. *Energy Research & Social Science* 18: 109–117.

Oliver, A. 2012. Markets and targets in the English national health service: Is there a role for behavioral economics? *Journal of Health Politics, Policy and Law* 37: 647–664.

Shelter. 2017. *Eviction of assured shorthold tenants*. Available at: http://england.shelter.org.uk/get_advice/eviction/eviction_of_private_tenants/eviction_of_assured_shorthold_tenants

Shove, E. 2010. Beyond the ABC: Climate change policy and theories of social change. *Environment and Planning A* 42: 1273–1285.

Vij, A., A. Carrel, and J.L. Walker. 2013. Incorporating the influence of latent modal preferences on travel mode choice behavior. *Transportation Research Part A: Policy and Practice* 54: 164–178.

Walker, G., N. Simcock, and R. Day. 2016. Necessary energy uses and a minimum standard of living in the United Kingdom: Energy justice or escalating expectations? *Energy Research & Social Science* 18: 129–138.

Caroline Mullen is a Senior Research Fellow in the Institute for Transport Studies at the University of Leeds, UK. Her research focuses on mobility and its implications for social, economic and environmental sustainability and justice, with an emphasis on travel needs, active travel, social and health equalities, and governance. Methodologically this involves the development of analytic frameworks for mobility planning, and social research methods, especially qualitative ones involving professionals and the public. Her research has informed transport strategies and planning (especially on cycling), and been used in collaboration with NGOs (active travel and mobility justice) and in public engagement.

Greg Marsden is Professor of Transport Governance in the Institute for Transport Studies at the University of Leeds, UK. His research interests relate to the why and how of policy making and in particular the interaction between different agents and agencies in the policy process. He works extensively on issues surrounding climate change, resilience and energy in the transport sector (and beyond). Recent research completed as part of the DEMAND Centre concerns mobility and understanding whether we can design solutions which improve well-being but do not inherently require greater mobility.

8

The Tenuous and Complex Relationship Between Flexible Working Practices and Travel Demand Reduction

Julian Burkinshaw

8.1 Introduction

Transport is a deeply complex and embedded socio-technical system. As such, any reduction in its dependence on fossil fuels, in order to address carbon emissions and the sustainability of related energy demand, will require a fundamental transition that goes beyond only technological changes (Watson 2012). The automobile commute is of central concern due to its contribution to carbon emissions, but has proven stubbornly resistant to established policy approaches (Cass and Faulconbridge 2016). Behaviour change interventions, which typically categorise the commute as a journey from home to work and back again (Department for Transport 2011), have long sought to shift the dominance of the private car for these journeys. However, their focus upon individual choices, through which behaviours are an outcome of attitudes (Shove 2010), does not acknowledge the underlying complexity of commuting. An alternative approach is therefore necessary to understand the complex

J. Burkinshaw (✉)
Institute for Transport Studies, University of Leeds, Leeds, UK

© The Author(s) 2018
A. Hui et al. (eds.), *Demanding Energy*, DOI 10.1007/978-3-319-61991-0_8

relationships between commuting and daily life, and their implications for changing patterns of travel.

This chapter builds upon growing interest in the use of 'social practice theory' (Shove et al. 2012; Spurling et al. 2013; Watson 2012) for considering transitions to lower carbon mobility, focusing in particular upon the interrelated practices of flexible working, of commuting and within the household. Discussions of 'flexible working' suggest that there is potential for both reducing the number of commuting journeys and shifting the timing of these journeys to address congestion and reduce the need for additional infrastructural investment. Yet this chapter argues that working practices themselves, and their links to other daily practices, significantly constrain the potential for changing patterns of travel.

The nature of work is changing, with less conventional temporal and spatial schedules having already emerged (Alexander et al. 2010). These are commonly referred to as 'flexible working practices'. Temporal and spatial flexibility in when, where and how work is done are key considerations within a number of accounts (Moen et al. 2008; Kelly et al. 2011; Schieman and Glavin 2008; Schieman and Young 2010; Baldock and Hadlow 2004; Yeraguntla and Bhat 2005), although a precise definition remains contested. Research into such changes has been extensive, with literature addressing effects on absenteeism (Van Ommeren and Gutiérrez-i-Puigarnau 2011) and implications for work-life balance (Wight and Raley 2009), amongst many topics. Pertinent discussions arise from research conducted into the effects of 'flexible working' on commuting; with Kim et al. (2015) in particular considering the efficacy of flexibility in both shifting the timing and reducing the number of daily commutes. As Cass and Faulconbridge (2016) have noted, shifting travel from the private car to more sustainable forms of mobility is extremely difficult. Flexible working practices are therefore considered a useful tool to help accomplish this shift by aiding in the daily sequencing, synchronisation and coordination of interrelated practices that may inhibit performances of lower carbon commute journeys (Cass and Faulconbridge 2016; Kim et al. 2015; Line et al. 2011; Pooley et al. 2011; Southerton 2009).

As such, flexible working practices are commonly perceived to help reduce demand for travel through both substitution and temporal shift.

Substitution involves reducing the need to travel, with home-working considered to be a major means of reducing carbon emissions associated with transport. Temporal shift on the other hand often revolves around two distinct approaches. The first is to 'flatten' frequently observed daily 'peaks', principally to reduce the burden upon the network during these times (mainly to combat congestion). The second is to encourage the use of lower carbon, more sustainable travel choices through leniency within arrival and departure times at work. This chapter questions the feasibility of these two dynamics. Through an analysis of in-depth, qualitative interviews this chapter will demonstrate how changing the overall levels of travel demand (and related energy demand) requires more than the flexibility of working practices because such flexibility can be constrained, with multiple practices affecting and shaping commutes. Flexible working thus does not relate to flexible commuting in any straightforward manner because of the constraints that working practices, and the coordination of these with other household practices, have upon when, where and why people must travel.

In analysing commuting, work and the household through theories of social practice, this chapter conceptualises them as interdependent practices that interact and combine to form bundles and complexes through which daily life is performed – and the sequencing, coordination and synchronisation of which shape patterns of demand for travel. Individual practices are defined by interdependent relations between three elements; *meanings, materials and competences*, referred to as the 'three-elements' model (Shove et al. 2012). This model has proved useful in understanding how practices are accomplished in time and space, whilst underlining the centrality of linkages between elements, practices and daily life.

What follows are discussions surrounding the constitutive elements from which practices are created, and how resultant practices of commuting, work and the household compete for time and space throughout the performance of daily life. These discussions are developed through consideration of practice sequencing, scheduling and coordination. Southerton (2006) explains that the temporal organisation of the day is constituted by both practices that have fixed and malleable positions within schedules. Consequently, the 'juggling' of activities by households is not necessarily a result of 'doing more'; rather certain tasks tend to fall together all at once, or have fixed institutional temporalities, such as working rhythms and school times. Attention is paid to how flexible working might help alleviate these sequencing pressures, in addition to what role practices within the household have in these pressures. Therefore, the purpose of this chapter is to discuss whether flexible working allows for the purported flexibilities in travel highlighted above, as the complexity of these interactions makes them worthy of greater investigation and analysis.

It was crucial for the purpose of this project to capture the intricacies of everyday practices and therefore semi-structured interviews were conducted with 29 participants from 6 professional groups: architects, academics, graphic designers, accountants, solicitors and university support staff. These professions were originally chosen because of their reputation as stereotypically 'flexible' and 'non-flexible' in terms of working arrangements. However, within the sample there ended up being little variation, with the majority of participants having access to flexible working, be it temporal, spatial or both. This suggests the prevalence of flexible working practices, and the process of "flexibilisation" (Garhammer 1995; Breedveld 1998) "whereby working times and locations are increasingly deregulated and scattered" (Southerton 2006: 438). The flexibility of work, however, does not necessarily correspond with flexibility of other practices, or travel. In this vein, the chapter will argue that flexibility of work does not necessarily equate to flexibility in travel for the commute, with limitations arising from working and other everyday practices. As such, flexibility within working practices is argued to be more supportive of the coordination and synchronisation of other practices, rather than of altering travel, with the prioritising of these other everyday practices having interesting consequences for commuting.

8.2   The Practice of Work and Its Implications for Travel to Work

Understanding the implications of working practices on patterns of travel to work requires a more thorough consideration of the dynamics that compel working at fixed times and spaces, even when flexibility may be possible. In doing so, this section of the chapter argues that flexibility of work does not necessarily equate to flexibility in travel for the commute. To illustrate this argument, the chapter draws upon examples that demonstrate the limitations for flexibility; notably the idea of 'core hours' around which flexibility is restricted, the constraint of meetings, the requirements of supervision and the way that Information and Communication Technologies (ICTs) cannot substitute wholly for the affordances of co-presence.

To begin, expectations and obligations arising from employers play a key role in determining both the demand and timing of travel. For example, the use of core hours (or office hours), the hours in which an employee has to be at work, plays a significant part in determining the timing of travel. In total, ten participants reported an expectation to be present in the workplace between specific times of the day, with core hours constraining over a third of participants in both time and space. Architect C, for example, reported core hours of:

> Half 8 until quarter past 5, so I get in between 8 and quarter past. That is observed by all of my colleagues too, so we are all in the office before 9. Then, maybe once or twice a week, probably once a week I will stay for an extra hour if needs be, but nine times out of ten I will leave at quarter past 5.

Adherence to these expectations and obligations influences the overall demand for travel as many participants were required to be present within the office on a daily basis. In addition, the timing of travel was influenced particularly through the hours in use being based upon societal norms of 9 to 5 working rhythms.

It is worth noting that whilst there is space for flexibility within working practices, it is typically limited whilst adhering to 'core' hour expectations and obligations introduced above. Pertinent here is that the conditioning of working practice rhythms and locales (e.g. core hours in the office) will invariably condition and frame the sequencing of practices in bundles and complexes (e.g. how working fits with commuting or household practices), as practices compete for finite resources in time and space.

In terms of temporally displacing travel, numerous participants exhibited occasions when travelling outside the 'peak', primarily to avoid the traffic encountered whilst travelling at those times. In order to enable them to do so, Accountant A and Solicitor A would work from home before travelling, allowing them to 'beat' the traffic. This did not negate their need to travel into the office, but encouraged them to travel outside of those peak hours, spreading the burden on the road network. Accountant A explains how:

> I start at 6am, so quite an early starter. Mostly that is to avoid traffic, if I'm honest. So I do some work at home and then travel in but I always try to avoid the rush hour, because life is too short to be stuck in the rush hour. Yeah so that is kind of a conscious decision to avoid the traffic.

This sentiment was echoed by Architect B, who advocated going into the office early and staying later to ensure 'focus time' to do some work on his own without interruption from his colleagues. A process evident throughout temporally flexible working practices was the concept of 'making up' time that had been lost or replaced due to starting later. Flexibility in this sense was reported by a number of participants as involving arriving later and consequently leaving later, or arriving earlier and then leaving earlier to enable the completion of personal tasks, such as attending a dentist appointment or doing some shopping. Over time, this has become more necessary and normal for a greater number of people as sequences of practices become increasingly squeezed both temporally and spatially (Southerton 2006). However, as we shall see later in this chapter, temporal flexibility further complicates the synchronisation of daily life, limiting opportunities to shift travel time thanks to adherence to socially significant and symbolic practices and elements, like those related to co-presence, supervision or the school run.

Client meetings, certainly for the architects, accountants, solicitors and graphic designers, play a key role in not only prompting travel to the office, but also travel between locations; be those site visits, court hearings, photo shoots, account discussions, and so on. Meetings are definitely a key component within each participant's working practices and rhythms, with coordination and synchronisation with others being of importance to the performance and reproduction of these working practices. Adhering to collective time structures like those of fixed working times, as Røpke and Christensen (2012) argue, makes coordination and synchronisation easier, whilst an increase in flexibility has meant an increase in the task of coordination. Although telephone communication and video conferencing technology use have risen in recent years, and are considered to be changing how and when people meet and communicate (Line et al. 2011), a large proportion of the meetings conducted by participants revolves around the co-presence and interaction of certain people in certain locations. This increases the propensity for time-space constraints within the working practices of participants and contributes towards the adherence of collective time structures. The prescriptive nature of client meetings therefore plays a role in determining the location and timing of meetings. Invariably, especially regarding site visits and out-of-(one's own) office visits, time and location are driven principally by a client (though often amicable for both parties). Interestingly, the split between participants who have meetings predominantly out-of-office and those who have predominantly in-office and internal meetings is almost 50-50. It is worthwhile noting that meetings in the sense spoken about above are probably 'typical' for participants, in contrast to more exceptional and infrequent meetings or conferences.

In addition to meetings, supervision requirements were considered a key component of working practices for several participants who therefore felt an obligation to be present within the office, be it to help resolve any issues that may arise or provide guidance and support to colleagues. These supervisory practices imposed temporal and spatial constraints upon the working practices of these individuals, limiting flexibility and demarcating the times at which they travel and spaces within which they worked. Accountant A provides a detailed account of why he feels the need to be in the office:

> I think part [of] coming to work is in all honesty having my colleague around, the social aspect of it. So it is coming in, prior to her being around, I would spend quite a lot more time at home... because she has been under training, she needs help, she needs guidance, and it is the social aspect of work isn't it. I think that supervision would be difficult from a distance,

because it is about sitting down at a screen, going through things, talking through technology I think would make that largely possible, but I don't think it is as effective as being there in the same room going through things next to each other, giving and receiving that immediate tutelage and feedback.

Additionally, Architect A, Accountants C and F, Graphic Designer A and Solicitor A all reported similar duties that imposed spatial constraints upon their working practices. However, the extent of these spatial constraints differed between participants. Architect A and Accountant F were able to manage their supervisory responsibilities effectively whilst away from the office thanks to their competence in using appropriate technologies and other materials. Furthermore, Accountant B described how available ICTs and related competencies have enabled supervisory responsibilities and commitments to be extended over greater distances and temporal boundaries. These competencies negate the need for co-presence in these circumstances; however, as the participant explains, the 'type' of work conducted during these face-to-face meetings via 'FaceTime' differs to that usually undertaken within the office when they are both present:

> Oh, there is FaceTime. So he goes away for a month over the festive period, we would FaceTime because he has an iPad and so do I. And that was the best way to do it, and that was checking in, every other day or so. It's checking, well it would be checking, you know are there any problems, you know I would be updating him on where I was. It is kind of like he is in the office, we just have to chat about anything that needs to be done. The only problem I think is if I need to help with more detailed things, detailed software and that sort of stuff, but that doesn't come around too often.

However, others did not find this process as complimentary or easy to manage. Accountant A for example felt that due to the software and technology, it would be easier to sit down at the computer screen than to use Skype or the telephone to work through any issues that have arisen.

While in some instances ICTs facilitate meetings while at a distance, in others co-presence is indispensable and therefore highly significant in generating temporal and spatial constraints for working. As Lyons (2013: 51) describes, co-presence offers "the prospect of a multi-sensory experience of encounter and exchange… matters of eye contact, body language and indeed smell can strongly colour the proceedings and outcomes of an activity." Possible difficulties may arise when trying to reduce travel for the journey to work as co-presence may be too important to particular working practices. Architect G elucidates that:

> I think it is working in an open plan office and being interactive a lot. It is not like you are in a cellular office cut off from everyone, you are chatting about with each other about the various jobs and bouncing ideas off each other and if you have got a query you don't know the answer to, you just ask someone. I find it a lot harder to work at home, so it is a lot better for us to come in and you know work together really. We can solve problems together as well. Things like communication side of it when I am at home I can log on to my computer and work, we are all on mobile phones now, so that isn't an issue, it is more that it is best to be with all the other guys to sort of, so we can question each other and get a bit of advice and things like that.

Co-presence, coupled with the confinement of working practices to within shared spaces and times, aids Architect G in creating a social setting within the workplace that is interactive, problem-solving and communicative. Therefore, rather than seeing a de-localisation of practices through an easing of time-space constraints (Røpke and Christensen 2012), the socially significant reproduction of these co-present working practices seemingly induces constraints within the working practices of these participants. Co-presence therefore seems to reaffirm Line et al.'s (2011) argument that assumptions should not be made as to the effectiveness of ICTs to influence levels of travel demand. Solicitor A explains how creating a 'central culture' or 'central body' from which to work is important, because doing so encourages a two-way mechanism that draws colleagues, clients and participants into the office, but also contributes towards the synchronisation and coordination of wider working rhythms. The spatially confined nature of a 'central culture' and 'central body' are reiterated by Shove et al. (2015) who contend that these institutionally determined patterns of timing and synchronisation relate to contingent forms of co-presence, coordination and power (amongst others). Therefore, the reproduction of this socially significant central working space, culture and associated rhythm coordinates a distinctive organisation of practices, which in turn affect travel practices for the commute.

In summary, it is evident that the dynamics of working, collaboration and the nature of shared understandings within working practices constrain opportunities for flexibility. Co-presence, meetings and supervision requirements are each significant, with coordination and synchronisation of activities, places and people of further importance in constraining when, where and how work is conducted. As a result, timing of and demand for travel to work are affected. ICT use can decouple activities such as meetings and supervision spatially; however, limitations to the success and strength of this decoupling were evident. In the next section, investigating the linkages between flexible working and other daily life practices will provide further insight into the complexity of the commute and the potential for reductions and shifts in travel demand for the journey to work.

8.3   Practices Within the Household and Their Implications for Travel to Work

To further understand the complexity of commuting, a consideration of how it interacts with work and other practices is necessary. Commuting is carefully coordinated and synchronised with other practices and people in such ways that altering this phenomena becomes difficult. Obligations and expectations associated with differing practices within the household impose similar limitations towards flexibility as those from working practices, likewise impacting upon times and demand for travel. It is argued henceforth that rather than supporting flexibility in travel for the commute, flexible working practices support the coordination and synchronisation of other everyday practices. Prioritising other practices such as the school run and exercise are shown to have interesting consequences for commuting.

Household responsibilities, and the practices that make up these responsibilities, play major roles in the synchronisation and coordination of daily life. Each participant in the sample described at least one responsibility that required their presence, be it in the home or elsewhere, at a certain time and location during the day, although not every day. These ranged from school drop-offs and pick-ups or elderly relative visits and care, to pet and animal care. Understanding how these responsibilities interact and incorporate into people's daily lives may be key to unpacking the 'commute', and helping to identify influences behind participant travel choices for the journey to and from work.

Importantly, around a third of the participants expressed that there was competition between household and work responsibilities on a daily basis, stressing a need for spatial separation between the two. In several cases, this separation increased the demand for travel into the office. Reasoning for the creation of this divide between home and work was evidenced from Architect E's past experience and knowledge of trying to work from home, finding that the household responsibilities bestowed upon him would almost override the need to do work.

School-run responsibilities shape daily travel for the majority of participants, with the need for coordination and synchronisation with other members of the household (namely partners) moulding travel choices further. Architect G describes how his commute has changed, principally through his school-run responsibilities, but also as a result of convenience, ease and job status:

Well I mean I used to catch the bus. I live in the suburbs, so it is pretty easy, the stop is a 5 minute walk to the office, so I used to use the bus a lot, but then becoming a director I decided that if the senior director has a car parking space then we should have one. So I drive in now. I have to do this as well because my kids have got older, so I am doing a lot of pick-ups and drop-offs in the morning and after school, so it is a lot easier. And also, because sometimes we do work a bit later it is a lot easier to drive than to have to go and find a bus in the evening.

Moreover, Architect G continued by explaining how coordination and synchronisation with both his partner's part-time work schedule and his children's after-school activities is essential, with at least one drop-off and pick-up per day a requirement. The need for synchronisation of these practices contributes towards the feeling that using the private car is a necessity, and convenience was a recurring theme throughout the interview. University Support Staff C elucidates that her daily schedule currently revolves around her daughter, with school-run responsibilities and after-school activities heavily dictating both work and journey times:

> Particularly on a Wednesday, I actually drive her to school, rather than drive her to the train station, because she has so many things to carry she is absolutely over-burdened with them… which means we leave home at 7, I drop her off and then drive to University, and I inevitably hit the traffic, you know, and it is quite a long journey, so I am not actually guaranteed being here for 9 o'clock when I do that…. So my travel is very much influenced by her and her activities, and if she has anything special, which she occasionally does, then you have to work around that.

The excerpts from Architect G and University Support Staff C above further illustrate that work times are not necessarily always focussed upon work demands, as the bundling of practices define and distort working rhythms. Processes of this nature are especially visible when dual income households recruit carers or use childcare facilities. Examples of these processes were provided by both University Support Staff D (USS D) and Architect E, where the utilisation of breakfast club for USS D "puts more pressure on me to get her somewhere and then into work." Although she could probably walk her daughter to school, her argument follows that "there is no point going backwards and forwards dropping her off and then getting the bus, I may as well drive her to school and then drive into University… when you are in the traffic you may as well just stay in it, you are stuck either way." When her mother is able to help with the childcare:

> My mum often picks her up from school, so you feel obliged that you don't want to be later than half 6 because my mum has had her from half 3 until half 6, and a 9 year old can't half create some noise! You have got to get back, there a little things like that in the back of your mind that when you are working late you make sure that you phone your partner to cover or one of us is in that sort of thing… so it is just balancing it between us really.

Interestingly and uniquely, Architect E demonstrates how a want for separation between work and home increases his demand for travel into the office. However, this is slightly negated by a variety of lower carbon, more sustainable modes used for the journey both to and from the office. These travel practices have been borne out of a need to synchronise the use of the family car with his partner, and to combine his commute with exercise and event training:

> At the moment my wife's on maternity leave, so she's at home most of the time, and so I usually drive my daughter to school, which is only a mile from our house. I'll then drive back home and have a choice of either running in, walking in, or cycling. Because generally I will leave the car at home and let my wife use that. Our office car is a pool car now, so generally anyone in the office can use it to go and visit sites or clients. The reason I have those three choices is because I do little events to keep myself exercising and healthy. So I would say I cycle three days a week to and from work... The other two days I'll either run both days in the morning and catch the bus back, or ride back if my bike is here, or I'll run to work and walk back. And that's because the event I am doing is carrying a rucksack over the Brecon beacons, so I'm kind of building up my miles and things like that, training.

Recruitment into these commute practices has been reinforced not only by the addition of exercise, but by the time-saving potential of these more-sustainable travel practices, and the use of an office pool car for journeys to meetings. Additionally, competences of both knowledge and experience gained from performing these practices further reinforce questions as to the value of using the private car for commuting:

> I hope to actually carry this on, because what I've found is, time wise, if I cycle it takes about 22 minutes along the tow path. If I run in I've done it in 35 minutes, and if I walk it can be about an hour and 5 minutes. When I have done it in the car at the time I set off, usually I will drop my daughter off at school at 10 past 8 at what we call breakfast club and then set off from there, so I can usually get in for 10 to 9, 9 o'clock. But if I am in a car, this morning it took me an hour to get in, which is pretty ridiculous!

University Support Staff D reiterates similar competencies, discussing that for the majority of the week she uses the bus to travel to and from work, though drives her daughter to school on Wednesdays. This entails booking a car parking space on University campus for the day – something that is relatively easy to do, but requires the competence and knowledge in order to apply for a permit, book the space and then access the campus via a private car. Furthermore, Architect E clearly demonstrates a number of these competences, given his ability to switch between modes seemingly effortlessly day by day and have the wherewithal to alter his route for safety, ease and enjoyment.

In summary, it is apparent that commuting is carefully coordinated and synchronised with other everyday practices. Bundling of the practices outlined above with those of the commute and work define and distort these practices, with work time especially not necessarily always focussed upon work demands. Similarities between work and practices within the household with regard to limited flexibility were discussed. Rather than supporting flexibility in travel for the commute, flexible working practices have been shown to encourage coordination and synchronisation with other activities, places and people. As a result of prioritising other everyday activities, the timing and demand for travel are affected, with spatial separation between work and home and the school run in particular having important consequences for commuting practices.

8.4   Conclusion

This chapter has argued that a deeper understanding of how practices synchronise, interact and coordinate is essential in order to unpack the dynamic interaction between how the daily commute, work and household practices are bundled. Discussion throughout the chapter has shown that the commute is far more complex than just the journey from home to work, and back again. It is a practice with diverse links to those of working, caring for family and leisure. It is the competition for time and space between these diverse practices that shapes the commute, with synchronisation and coordination of responsibilities within daily life of paramount importance.

Upon reflection, it can be argued that flexibility within working practices is minimal, with little disparity found between professions, highlighting the saliency of temporal sequences and collective patterns of action. Examples of both shifting demand temporally and reducing demand spatially were introduced. However, flexible working practices could be construed as having limited effects upon the timing of travel and overall travel demand for work, therefore putting into question the efficacy of these policies to reduce energy demand within transport. Whilst flexible working arrangements are not difficult to facilitate, the extent of their use was minimal. It is therefore perhaps interesting to consider that the minimal effects of flexibility have been normalised. Doing so raises questions regarding the future scope of flexible working practices, with the structure of work and household responsibilities being dominant features within the performance and organisation of everyday life practices. Limitations relating to the temporal and spatial flexibility of these practices, and the bundles they engender, could continue to restrict shifts in the timing of travel or reductions in overall travel demand to work.

Acknowledgements This work was supported by the Engineering and Physical Sciences Research Council [grant number EP/K011723/1] as part of the RCUK Energy Programme and by EDF as part of the R&D ECLEER Programme.

Bibliography

Alexander, B., M. Dijst, and D. Ettema. 2010. Working from 9 to 6? An analysis of in-home and out-of-home working schedules. *Transportation* 37: 505–523.

Baldock, J., and J. Hadlow. 2004. Managing the family: Productivity, scheduling and the male veto. *Social Policy & Administration* 38: 706–720.

Breedveld, K. 1998. The double myth of flexibilization trends in scattered work hours, and differences in time-sovereignty. *Time & Society* 7: 129–143.

Cass, N., and J. Faulconbridge. 2016. Commuting practices: New insights into modal shift from theories of social practice. *Transport Policy* 45: 1–14.

Department for Transport. 2011. *Commuting and business travel*. Available at: https://www.gov.uk/government/uploads/system/uploads/attachment_data/file/230553/Commuting_and_business_travel_factsheet___April_2011.pdf

Garhammer, M. 1995. Changes in working hours in Germany the resulting impact on everyday life. *Time & Society* 4: 167–203.

Kelly, E.L., P. Moen, and E. Tranby. 2011. Changing workplaces to reduce work-family conflict schedule control in a white-collar organization. *American Sociological Review* 76: 265–290.

Kim, S.-N., S. Choo, and P.L. Mokhtarian. 2015. Home-based telecommuting and intra-household interactions in work and non-work travel: A seemingly unrelated censored regression approach. *Transportation Research Part A: Policy and Practice* 80: 197–214.

Line, T., J. Jain, and G. Lyons. 2011. The role of ICTs in everyday mobile lives. *Journal of Transport Geography* 19: 1490–1499.

Lyons, G. 2013. Business travel—The social practices surrounding meetings. *Research in Transportation Business & Management* 9: 50–57.

Moen, P., E. Kelly, and Q. Huang. 2008. Work, family and life-course fit: Does control over work time matter? *Journal of Vocational Behavior* 73: 414–425.

Pooley, C.G., D. Horton, G. Scheldeman, et al. 2011. Household decision-making for everyday travel: A case study of walking and cycling in Lancaster (UK). *Journal of Transport Geography* 19: 1601–1607.

Røpke, I., and T.H. Christensen. 2012. Energy impacts of ICT – Insights from an everyday life perspective. *Telematics and Informatics* 29: 348–361.

Schieman, S., and P. Glavin. 2008. Trouble at the border: Gender, flexibility at work, and the work-home interface. *Social Problems* 55: 590.

Schieman, S., and M. Young. 2010. Is there a downside to schedule control for the work-family interface? *Journal of Family Issues* 31: 1391–1414.

Shove, E. 2010. Beyond the ABC: Climate change policy and theories of social change. *Environment and Planning A* 42: 1273–1285.

Shove, E., M. Pantzar, and M. Watson. 2012. *The dynamics of social practice: Everyday life and how it changes.* London: Sage.

Shove, E., M. Watson, and N. Spurling. 2015. Conceptualizing connections energy demand, infrastructures and social practices. *European Journal of Social Theory* 18: 274–287.

Southerton, D. 2006. Analysing the temporal organization of daily life: Social constraints, practices and their allocation. *Sociology* 40: 435–454.

———. 2009. Re-ordering temporal rhythms: Coordinating daily practices in the UK in 1937 and 2000. In *Time, consumption and everyday life: Practice, materiality and culture,* 49–63. Oxford/New York: Berg.

Spurling, N., A. McMeekin, E. Shove, et al. 2013. *Interventions in practice: Re-framing policy approaches to consumer behaviour.* Manchester: Sustainable Practices Research Group. Available at: http://www.sprg.ac.uk/uploads/sprg-report-sept-2013.pdf. Accessed 15 May 2017.

Van Ommeren, J.N., and E. Gutiérrez-i-Puigarnau. 2011. Are workers with a long commute less productive? An empirical analysis of absenteeism. *Regional Science and Urban Economics* 41: 1–8.

Watson, M. 2012. How theories of practice can inform transition to a decarbonised transport system. *Journal of Transport Geography* 24: 488–496.

Wight, V.R., and S.B. Raley. 2009. When home becomes work: Work and family time among workers at home. *Social Indicators Research* 93: 197–202.

Yeraguntla, A., and C. Bhat. 2005. Classification taxonomy and empirical analysis of work arrangements. *Transportation Research Record: Journal of the Transportation Research Board* 1926: 233–241.

Julian Burkinshaw is a PhD Researcher in the DEMAND Centre and the Institute for Transport Studies at the University of Leeds, UK. He received his bachelor degree in Geography from the University of Salford in 2012. His doctoral research focuses on flexible working practices and what role these may have in shifting and reducing travel demand for the journey to work. It involves investigations into interactions between work, travel and the household in order to further understand the complex phenomena of commuting. This approach will provide insight into potential energy demand reduction within transport.

9

Leisure Travel and the Time of Later Life

Rosie Day, Russell Hitchings, Emmet Fox, Susan Venn, and Julia F. Hibbert

9.1 Introduction

The tourism industry in developed countries worldwide is anticipating a rapid increase in older travellers (European Commission 2014; Tarlow 2007; ABC 2015). A large part of the reason for this is demographic change: increasing proportions of the population are older (Harper 2014), and there are growing numbers of retired people, especially as the

R. Day (✉)
University of Birmingham, Birmingham, UK

R. Hitchings
University College London, London, UK

E. Fox
Anglia Ruskin University, Cambridge, UK

S. Venn
University of Surrey, Guildford, UK

J.F. Hibbert
Bournemouth University, Poole, UK

© The Author(s) 2018
A. Hui et al. (eds.), *Demanding Energy*, DOI 10.1007/978-3-319-61991-0_9

large 'baby boom' generation born between 1946 and 1964 in Europe, the US, Canada, and Australia are now reaching retirement age. In the UK, the number of people of state pension age or older is projected to rise from 12.3 million in 2015 to 14 million in 2030 and 16.6 million by 2040 (Pensions Policy Institute 2013). On average, they also have increasing financial resources (*The Economist* 2016) and in the UK retirees are also now able to draw down lump sums of money from their pension savings in a way not possible before (Urzi Brancati and Franklin 2015). In addition, the baby boomer generation are much discussed as being avid consumers, in ways which they are likely to carry forward into retirement (Street and Cossman 2006; Gilleard and Higgs 2005). They are also a generation that has experienced a great expansion in opportunities for leisure travel over their lifetimes, especially in overseas travel, through a series of changes to infrastructure and service provision, from the development of access to airports from the late 1950s through the spread of affordable package holidays in the 1980s, to the arrival of low cost airlines and expansion of regional airports. As such, they are likely to have higher expectations for travel than earlier cohorts of older people did, as will following cohorts.

If such a boom in senior travel does come about, there will be significant implications for derived energy demand and associated carbon emissions. Such predictions of a large increase are however generally based on demographic trends and rest on assumptions about what older people might do under changing circumstances such as increasing incomes. Quantitative data on actual travel is a little mixed for the UK regarding whether there is at the moment a marked increase in travel among retirees. Data from the UK's International Passenger Survey indicates people aged over 65 making up an increasing proportion of travellers by air—9.8 % of their respondents in 2012 compared to 7.2 % in 2001. Data from the UK Living Costs and Food Survey 2001–2012 on numbers of flights taken annually shows a rising trend for 65–74 year olds from 2001 to 2007, followed by a fall from 2007 to 2011, and some recovery 2011–2012, while for those aged 75+ the trend is essentially flat over these years, as well as much lower in terms of number of flights than for other ages.

In order to understand more clearly how travel among older people is happening and likely to evolve, we need to look beyond such statistical trends to understand how demand for longer distance mobility in later life is made. Older age as a life course stage has some specificities that are likely to impact on travel; this has been noted up to a point in work on mobility biographies (Scheiner 2007; Müggenburg et al. 2015) which has, for example, highlighted retirement as a key event which is likely to change patterns and modes of mobility; that literature however has focused only on everyday mobility and on what people are observed to be doing. We argue that we need to take more account of not just one key moment of change, but of how the move into and through retirement and later life is thought about and experienced, in order to understand how travel practices continue, evolve or are remade in changing circumstances in older age. Here, we are also more interested in the less frequent and longer distance leisure mobility that is usually termed 'holidays', or 'travel'; this is of interest as it is the most resource intensive form of mobility and unlike everyday mobility, it is predicted to be particularly on the rise for older people.

In order to consider in more depth the dynamics shaping such energy-intensive travel demand in later life, we discuss material from a qualitative study of three cohorts of people aged over 50 in the UK. Based on an inductive analysis of interview discussions, we consider some different ways in which time is implicated in how leisure travel is practised in later life: time in terms of how it is constructed, how older age is experienced temporally, and how circumstances organise and use up time in particular ways. Drawing on discussion of different forms of temporality by Szerszynski (2002) we consider how retirement is viewed in the kairological sense of time, given meaning as a particular episode of life which is the 'right' time for certain things; the linear temporal progression of embodied older age; and the time of older age as a resource which is called on by others, and increasingly colonised by caring activities and responsibilities. More nuanced attention to time and later life in this way we argue provides for a more sophisticated understanding of the development of leisure travel among older cohorts and its likely trajectory.

9.2   Study Design

We carried out an interview-based study with 60 households located in London and Birmingham, UK. Twenty households were in each of three age cohorts: the youngest (cohort 1) were aged 50–55 and yet to retire (based on the main earner); cohort 2 were aged 60–69 and were less than 7 years retired; and cohort 3 were aged 75+ and retired for 10 or more years. This structure was in order to capture a sense of different age cohorts' expectations regarding retirement travel, whether that was still ahead of them, or whether they were some way into retirement. Within each age cohort, households were split equally between two social class categories, with social class being categorised based on the main occupation or previous main occupation of the main wage earner, in order to account for the differences in consumption patterns likely to be associated with membership of different socio-economic groupings. It is important to note though that social class grouping does not correspond exactly with income grouping, especially in retirees. The 'higher social class' (hereafter HSC) group comprised households in social classes A or B, and the 'lower social class' (hereafter LSC) group, households in classes C2, D or E. Households in each cohort also included couples and people living alone, both male and female.

Each of the 60 households undertook two in-depth, qualitative interviews over the course of several weeks. These discussed past and present leisure travel of various forms and distances, how it was organised and how they would likely organise it in the future. The idea was to step back from predictions such as those with which we started this chapter, to explore how leisure travel had evolved in their lives and how that intersected with different processes associated with ageing, both real and anticipated. Thematic analysis based on themes derived from guiding research questions and more inductively from iterative, collaborative reading of transcripts helped to draw out the findings. Here, we discuss three aspects of time associated with older age as our participants thought about and experienced this stage of their lives, and the implications of these for their longer distance travel.

9.3 Retirement as a Life Episode: Freedom and Self-Fulfilment

Retirement, understood as the end of paid employment, is generally held to be a distinct temporal phase of life that most often starts in people's 60s, notwithstanding recent and ongoing changes to the age of state pension eligibility in the UK. With the cessation of employment and its associated routines, time has the potential to become less structured, and more available for leisure. This can be read as a move from the domination of clock-time that organises labour, into a distinct segment of life with a different meaning and purpose, and temporal rhythm (Macnaghten and Urry 1998; Szerszynski 2002). For our youngest cohort of participants, those yet to retire, this freeing up of time was something they saw themselves making the most of for travel:

> I would like to travel when I have more time. I think time is more of the issue really. Doing four, six week trips....I'd like to travel as much as I can. (Lawrence, 50–55, higher social class)

> I do think once I retire, I would really hope to be travelling quite a lot actually. (Luisa, 50–55, LSC)

Cohort 1 participants had on the whole quite extensive hopes and dreams regarding travel once they retired, involving diverse, global destinations and sometimes being away for long periods; there was some mention of 'pent up demand'. Cohort 2 as the relatively recent retirees appeared to some extent to be putting these dreams into practice. Rosemary and Simon from London, (LSC group) when interviewed in the spring were planning for a cruise from Arabia to India and another down the Danube, an autumn trip to Turkey and a visit to their time-share in Cyprus, plus a Kenyan safari in the foreseeable future. Others intended to visit Iceland, Cambodia, California, the Silk Road, New Zealand—although it should also be noted that the amount and frequency of their actual travel varied quite considerably.

The kairological or qualitative construction of retirement as a time of freedom, as noted also in other work from beyond the UK (Weiss 2005; Hunter et al. 2007) rests on its contrast with pre-retirement life, where time is more constrained by duties and commitments associated with family, and especially working life, which for some may not be particularly fulfilling. Travel then could be rationalised as a reward justifiably reaped, as part of an idealised life stage that in part might serve to make up for the grind of earlier years, where time was for labour and duty:

> But I think well I've had my working life. I've provided and I've hardly had any days off for the holiday and what have you. I think I perhaps deserve something. But maybe that's being a bit selfish. (Brian and Christine, 60–69 HSC)

Whilst current holidays for those in their 50s were something to be fitted in, and largely about relief from work and domestic routines, and quality time with family, travel for the retirees was about having experiences, expanding horizons, pleasure and learning, much more based around self-actualisation. It was as a generality felt to be a good thing for older people, connected also with internalised ideas about the necessity of staying active and avoiding decline (Katz 2000) and a common refrain was that everyone should travel as much as they want in retirement if they can afford, financially, to do it.

Among the older cohort, those aged 75+, travel had also featured in hopes for retirement at the time for many, but to a lesser extent than for other age groups in our sample and also more reported by the HSC group than the lower. Overall it appeared that although they subscribed to the idea of travel in later life being a good thing, this cohort's expectations for their own retirement travel had been somewhat less expansive, involving fewer long distance trips, and some had not planned to travel much at all. The reasons for this are quite complex, and in large part, we argue, due to the norms around travel distance in their earlier lives that they took into retirement, linked to the development of travel infrastructure (see Fox et al. 2017). As such, we can see longer distance retirement travel becoming normalised over time. This is aided by the travel industry which, hoping to profit from the growing potential market of older people with financial resources (Tsiotsou et al. 2010; Muller and O'Cass 2001), is increasingly marketing ambitious travel as part of an idealised retirement.

It must also be noted though, and it was by our participants, that another important feature of retirement was the restructuring of finances and the move to relying on annuities and savings which in general provide a lower income than when working and are more susceptible to stock market and interest rate fluctuations. The reality of retirement income once they got there meant that some people were not quite as able to make their dreams concrete as they would have liked:

> I would like to do more [travel] but obviously being on a fairly tight budget, I'm going to be pretty fairly limited … I mean there's loads that I would like to do but I've got to be a little bit sort of choosy. (Gloria, 60–69, HSC)

The recent financial downturn and its ramifications had amplified this and meant that quite a lot of participants had experienced a lowering of their income and many of cohort 1 participants were aware that their pensions were going to be worth less than they had expected. So a number of households found themselves unable to travel as much as they hoped to and even the better off ones were aware that finances would be a constraining factor on their ambitions. The idea of 'blowing the pension' on travel (as has been a topic of concerned speculation surrounding recent regulatory changes in the UK) was uncommon. Although some households said that they were spending more on travel whilst their health and finances allowed, the more common view was that it was necessary to spend responsibly, with an eye to future needs.

Retirement then was related to in an ideal sense as a specific segment of life (Szerszynski 2002) where leisure and enjoyment were deserved and self-actualisation came to the fore, and significant travel was unproblematically part of this to the extent that one could responsibly afford it. The younger two cohorts, together comprising the baby boomer generation, had more ambitious travel plans than those aged 75+ had done, echoing the wider expectations of their generation and their consumption habits. So far, this is much as predicted at the start of this chapter; however, the picture becomes more complicated as other time-related aspects of later life come into play.

9.4 Physical Ageing Anticipated and Lived in Linear Time

In contrast to the relative suddenness of retirement, signalling a distinct phase of life, the experience of bodily ageing is usually more gradual, though also sometimes it does involve more abrupt shifts and episodes of illness. Its temporality is linear, similar to that described by Szerszynski (2002) as linked to the development of capacities and skills, but in this case rather more characterised by decreasing capacities rather than increasing.

Among our participants, health and mobility concerns related to ageing were a major influence on their travel, even when they weren't being directly experienced. They were not, however, a straightforward limitation. Some of our interviewees, particularly among the younger two cohorts, talked about how an awareness of increasing bodily limitations and a heightened sense of the possibility of health problems, even death, striking at any point, led them to step up their travel activity at least for a while, in a kind of 'do it while you can' phase.

> Evan: …because my mum dropped dead in her 60s but that period between early 60s and late 60s where you are fit and well enough to travel, really serious travel, I'm not talking about you know popping down to Bournemouth but sort of proper, proper international travel, there's really that window of about ten years – where before your health or your insurance or you're just not so mobile.
>
> Ruth: And you don't know what's going to happen do you? (Ruth and Evan, 50–55, HSC)

Anticipated ageing then lent a sense of quality retirement time, as described in the earlier section, being finite, possibly even short, and to be made the most of.

For quite a lot of our households, and more so as the age of them increased, the current health and mobility of one or more of them was placing clear limits on their more distant travel (and sometimes on their nearer). Some degree of mobility limitation was quite common in the middle cohort of people in their 60s, and both more common and more significant in the older cohort of people over 75. Specific health and mobility issues might be the limiting factor, but also just general energy levels and inclination led some to choose closer destinations for their holidays.

> I enjoyed that [going on further holidays/cruises]; I mean I'd love to be able to do it still but you know you can't so you have to accept that…. I think since my back's been so bad I can't do it, I suppose it's about 10 years. (Agnes, 75+, LSC)

> I don't know, I won't go abroad again, no I don't think so, I don't think I'll fly again… I just don't fancy it any more…. No, I've done all that, I can't imagine myself doing that again. You've got to take everything into consideration, my age and things like that, and then of course don't forget the insurance is a lot more so now, especially. (Bob, 75+, LSC)

Mobility and health limitations did appear to be more frequent and more severe in our LSC groups and the difference was most clear in the 75+ cohort. Although our sample size was too small to draw wider conclusions on such patterns, this would be congruent with quite apparent inequalities at national level in healthy life expectancies (Office for National Statistics 2015), affecting their later life opportunities.

As well as their own health, the health of travelling companions (outside of the household) was relevant, especially for the single householders. Even if they were personally healthy and mobile, their companion might be more limited and this would have a bearing on the types of travel they could do. Travel for some was also interrupted or curtailed by bereavement and the loss of a travel companion, either a partner or a friend. Whilst some people were happy to travel solo, or had found suitable providers that catered for their needs as solo travellers, others were not comfortable with travelling alone. Single person supplements were also noted as increasing financial strain, but hard to avoid.

Sometimes, for those who could continue to afford it, mobility limits might mean that rather than further travel being limited or stopped, the mode of travel or type of holiday was changed, for example a switch to cruises, even among those who previously would not have chosen cruises for reasons of them being restrictive, or associating them with the 'elderly'.

> If you got older you see that might change because then a cruise is easier as you get older because you're sort of taken in hand, and everything is pro-vided for you… (Matthew and Joy, 60–69, LSC)

In addition, and prompted as well by legislation such as the UK Equalities Act 2010 (HM Government 2010) innovations such as wheel-chair assistance at airports were enabling some of our participants carry on with long distance travel. Provision among travel and tourism service providers for disability and other health issues seems at yet relatively lim-ited though (Buhalis et al. 2012). Providing for specific needs of older and disabled customers is not something that those in the tourism indus-try have always found easy to handle – is it better to downplay ageing when marketing to the old because no one likes to think about these issues or is it better to recognise them and show how they can be man-aged? (Tsiotsou et al. 2010). At the moment then, a decline in mobility is likely to restrict the amount and type of leisure travel that people do. This bodily reality of ageing was, when encountered by our participants, acknowledged and pragmatically worked around.

The embodied linear temporality of later life therefore served both as an impetus to travel more, at least for a period, in order to grasp the imagined, idealised retirement whilst it was possible, and then a signifi-cant constraint on travel for many that necessitated some renegotiation of their travel related hopes and expectations. A further counter to the hedonistic, individual goal-seeking ideal of retirement came about through the demands of relationships, to which we now turn.

9.5   Evolving Relationships and the Temporal (Re)distribution of Care

The third aspect of older age and time that we discuss here is that condi-tioned by the ongoing evolution of close relationships and especially the changing configurations of caring roles and responsibilities as people move towards and through older age. This evolution and its significance for travel were quite apparent looking across our participants in the different age cohorts. Among our youngest cohort of pre-retirement households in their 50s, some still had children young enough not to be left alone while they holidayed, so children and their needs were a major factor in the amount and style of travel that they did. As such, some of them were looking forward to being able to travel differently and more flexibly when they were freed from the constraints of school holidays and/or destinations that suited their children.

> I would like the idea that potentially in five years' time perhaps my youngest might be independent and going on a school holiday and my eldest is away and I could go, I could go somewhere, depending on the level of fitness. Yeah, not mountain climbing, but doing something like going sailing, windsurfing, those things. (Jenny, 50–55, HSC)

This moving on from parenting roles was part of the younger cohort's imagined retirement-as-freedom discussed earlier. Yet, the experience of the older two cohorts showed that family responsibilities didn't necessarily fall away, but rather they often assumed a different shape, sometimes unexpectedly.

Quite a lot of our households had grandchildren, and were to varying degrees involved in their care. This inevitably took up a lot of their time, leaving less for their own leisure, and might also tie them into a routine of expected care provision, providing a further constraint on the flexibility to travel at any time or stay away for long. Grandparenting roles also involved spending money that might otherwise have been devoted to individual travel.

> I think I would probably have more free time if, if I didn't look after grandchildren so that might have affected the holidays rather than, it's not retirement itself it's the fact that my time's filled up doing different things. (Penny, 60–69, LSC)

Even more commonly though, our research participants were caring for other older people. Some of our participants in their 50s and 60s were relied on by parents needing support, and then increasingly with increasing age, they were caring for partners, and others in their own age group. Connie is an example:

> I've got a brother over the road, he's got terminal cancer, he had an operation on 13th March last year and from then until now it's hospital, doctors, clinics, blood bank, God knows what, that is my life now, since last March. So I've got my own husband [who she is a carer for] which isn't a problem with him really, he's alright. It's him over there, trying to look after him. (Connie, 75+, LSC)

Miranda meanwhile (75+, LSC) had a husband in a care home, with Parkinson's disease and dementia. Although she no longer fully cared for him herself, she visited him 6 days a week. Both Connie and Miranda demonstrate how such responsibilities can put severe restrictions on the ability to travel far at all, preventing even short holidays away (see also Gladwell and Bedini 2004).

However, as a counter-current in terms of travel impetus, relationships in retirement also provided reasons to travel. Even though grandparenting might keep people closer to home much of the time, that role could also invoke travel, as more than one participant planned specific holidays with their grandchildren as a way of spending time with them. Apart from that, younger generations of their families had often moved away, and the desire to see them in person provided a strong imperative to travel longer distances than they perhaps might have done otherwise. While much of the talk of visiting family related to going to other parts of the UK, many participants had children, relations or close friends who had moved overseas either permanently or temporarily, which meant that leisure trips were taken to visit them.

> My brother works in the United States, he organised a holiday for his family and it's his son who's the nephew I go and see in Manchester so he was organising for them to go out and see him in Michigan. He invited me along so I did that. So that was two weeks in Michigan in the summer and then previously over Christmas not this Christmas but [the one before] my sister who lives in Australia was going to Bali for Christmas and invited me along. So I went to Bali for Christmas and then she went back to one part of Australia but I've got a friend who was living in Australia so I went to see him. (James, 60–69, HSC)

Often this led to or enabled more travel than otherwise because staying with family and friends meant that it was cheaper, as well as being a higher priority. For others though it meant that their travel opportunities were prescribed—for example, Irene (60–69, LSC) spent all her travel budget on flying to California twice a year to see her daughter and grandchildren, and she wasn't able to go on other holidays that she might otherwise have wanted to, in less distant locations (she found the long haul travel difficult).

Travel then needed to work alongside relationships and care responsibilities that reconfigured as our participants moved beyond mid-life. Sometimes, in these times of increasingly globally dispersed families, this could mean more travel and there were cases of people 'spending the kids' inheritance'—another stereotype of self-indulgent baby-boomer travellers – by actually going to see the kids. For others though, although they might have nurtured dreams of wider travel in retirement, the needs of their loved ones consumed their time and meant that they felt they needed to stay closer to home.

9.6   Discussion

We have highlighted three aspects of time in older age: the kairological time of retirement as a time for self-fulfilment, the linear temporal progression of embodied ageing, and the consumption of time by reconfigured caring responsibilities, all of which act, together and in competition, to shape the demand among older people for longer distance leisure travel. The qualitative design of our study allowed insights not possible through the more usual assessments of demographic patterns and of travel surveys in this field, and contributed to addressing a highlighted lack of such work (Sedgley et al. 2011). The cohort design of our study allowed us to consider, although not be conclusive about, the temporality of the life course by considering people at different stages of retirement, and also to reflect on potential change between cohorts in the ways in which they approached and experienced this life stage.

We saw retirement imagined as a time of freedom and self-fulfilment, for the pursuit of individual goals; and for all three age cohorts, travel was seen as a potentially important part of this, but for the younger two cohorts, both of the baby boomer generation, their expectations were greater, and their norms of travel involved longer distances and more 'exotic' destinations (see discussion in Fox et al. 2017). The financial aspects of retirement though were not always fully supportive of their desires, and generally, the dreams were offset with notions of financial responsibility where awareness of the other aspects of ageing crept in. The embodied progression of ageing acted as a prompt to speed up the execution of travel plans but later to signal the need to renegotiate the specifics of personal travel goals. The demands of relationships challenged the individualistic notions of idealised retirement, pulling people back into less self-focused activities and often recolonising their time with duties and routines of care for others.

How 'senior travel' and associated energy demand and carbon emissions evolve, in the UK context but also more widely, depends not just on numbers of older people and their financial resources but on how the dynamic between these aspects of older age plays out. The idealised, imagined retirement as a time for deserved, self-centred pleasures is made much of in popular discourse and in marketing, but in understanding where future travel demand among older people is likely to go, other aspects are in danger of being overlooked. Financial resources are of course required for travel and these are subject to the vicissitudes of national and global economies, as pensions depend heavily on interest rates. Physical ageing will continue to happen and although disability-free life expectancy is on an upward trend in the UK, healthy life expectancy currently is not (Office for National Statistics 2014); and here again there is much diversity and inequality (Office for National Statistics 2015). In this regard, the extent of future senior travel may depend a lot on the extent to which service providers evolve to cater for older age and specific health and mobility related needs and thus to extend the 'travel while you can' phase of an ideal retirement. Currently such provision is apparently quite limited (Buhalis et al. 2012), but there are calls for this to change, for reasons of social equality but also in order for the industry to realise the potential of the senior and disabled market (Alén et al. 2012; Özogul and Baran 2016). The colonisation of people's later years with caring responsibilities is also responsive to economic and political currents: recent large increases in the number of older people acting as carers for other older people (Age UK 2016) as well as grandchildren and others, is undoubtedly connected to ever-reducing state provision of adult social care, child benefits and other kinds of welfare support. This reinforces the point made elsewhere that energy demand is configured by policy in all kinds of domains, not only that which is considered as 'energy policy' (Cox et al. 2016).

In configuring energy demand and emissions, it is not only the frequency of travel that will matter but choices about distance and mode. The effects of some age-related choices may be surprising, for example per capita emissions associated with long distance cruises, perhaps chosen as an easier form of travel, can be as high or higher than those associated with long haul flying (Howitt et al. 2010). As much as affecting overall energy consumption through travel then, population ageing is likely to affect where and in which sector this demand occurs. While broadly speaking the effect of population ageing and greater longevity may be to increase desire for long distance leisure travel, the patterns of energy demand that arise will be the outcome of complex dynamics, as older people negotiate the tension between their retirement fantasies and the evolving needs of their own bodies, their loved ones and the imperative to provide for their own future welfare in changing circumstances.

Acknowledgements This work was supported by the Engineering and Physical Sciences Research Council [grant number EP/K011723/1] as part of the RCUK Energy Programme and by EDF as part of the R&D ECLEER Programme.

Bibliography

ABC. 2015. *Increase in elderly travellers poses new challenges for airline industry.* Available at: http://www.abc.net.au/radionational/programs/breakfast/increase-in-elderly-travellers-poses-new-challenge/6565668

Age UK. 2016. *Invisible but invaluable army of carers save state billions.* Available at: http://www.ageuk.org.uk/richmonduponthames/news--campaigns/invisible-but-invaluable-army-of-carers-save-state-billions/

Alén, E., N. Losada, and T. Domínguez. 2012. New opportunities for the tourism market: Senior tourism and accessible tourism. In *Visions for global tourism industry – Creating and sustaining competitive strategies*, ed. M. Kasimoğlu. Rijeka: InTech.

Buhalis, D., S. Darcy, and I. Ambrose, eds. 2012. *Best practice in accessible tourism: Inclusion, disability, ageing population and tourism*. Bristol: Channel View Publications.

Cox, E., S. Royston, and J. Selby. 2016. *The impacts of non-energy policies on the energy system: A scoping paper*. London: UKERC. Available at: http://www.ukerc.ac.uk/asset/1B9BBB2F-B98C-4250-BEE5DE0F253EAD91/

European Commission, Enterprise and Industry Directorate-General. 2014. *Europe, the best destination for seniors: Facilitating cooperation mechanisms to increase senior tourist's travels within Europe and from third countries in the low and medium seasons – Experts draft report*. Available at: http://ec.europa.eu/DocsRoom/documents/5977/attachments/1/translations/en/renditions/native

Fox, E., H. Russell, R. Day, et al. 2017. Demanding distances in later life leisure travel. *Geoforum* 82: 102–111.

Gilleard, C., and P. Higgs. 2005. *Contexts of ageing: Class, cohort and community*. Cambridge: Polity.

Gladwell, N.J., and L.A. Bedini. 2004. In search of lost leisure: The impact of caregiving on leisure travel. *Tourism Management* 25: 685–693.

Harper, S. 2014. Economic and social implications of aging societies. *Science* 346: 587–591.

HM Government. 2010. Equality act 2010.

Howitt, O.J., V.G. Revol, I.J. Smith, et al. 2010. Carbon emissions from international cruise ship passengers' travel to and from New Zealand. *Energy Policy* 38: 2552–2560.

Hunter, W., W. Wang, and A. Worsley. 2007. Retirement planning and expectations of Australian babyboomers. *Annals of the New York Academy of Sciences* 1114: 267–278.

Katz, S. 2000. Busy bodies: Activity, aging, and the management of everyday life. *Journal of Aging Studies* 14: 135–152.

Macnaghten, P., and J. Urry. 1998. *Contested natures*. London: Sage.

Müggenburg, H., A. Busch-Geertsema, and M. Lanzendorf. 2015. Mobility biographies: A review of achievements and challenges of the mobility biographies approach and a framework for further research. *Journal of Transport Geography* 46: 151–163.

Muller, T.E., and A. O'Cass. 2001. Targeting the young at heart: Seeing senior vacationers the way they see themselves. *Journal of Vacation Marketing* 7: 285–301.

Office for National Statistics. 2014. *Health expectancies in the United Kingdom, Great Britain, England, Wales, Scotland & Northern Ireland.* Available at: https://www.ons.gov.uk/peoplepopulationandcommunity/healthandsocial-care/healthandlifeexpectancies/datasets/healthexpectanciesintheunited kingdomgreatbritainenglandwalesscotlandnorthernireland

———. 2015. *Inequality in healthy life expectancy at birth by national deciles of area deprivation: England, 2011–2013.* Available at: http://webar-chive.nationalarchives.gov.uk/20160114055601/http://www.ons.gov.uk/ons/rel/disabilityand-health-measurement/inequality-in-healthy-life-expectancy-at-birth-by-national-deciles-of-area-deprivation-eng-land/index.html

Özogul, G., and G.G. Baran. 2016. Accessible tourism: The golden key in the future for the specialized travel agencies. *Journal of Tourism Futures* 2: 79–87.

Pensions Policy Institute. 2013. *Demographics.* Available at: http://www.pen-sionspolicyinstitute.org.uk/pension-facts/pension-facts-tables/table-1-demographics

Scheiner, J. 2007. Mobility biographies: Elements of a biographical theory of travel demand. *Erdkunde* 62: 161–173.

Sedgley, D., A. Pritchard, and N. Morgan. 2011. Tourism and ageing: A trans-formative research agenda. *Annals of Tourism Research* 38: 422–436.

Street, D., and J.S. Cossman. 2006. Greatest generation or greedy geezers? Social spending preferences and the elderly. *Social Problems* 53: 75–96.

Szerszynski, B. 2002. Wild times and domesticated times: The temporalities of environmental lifestyles and politics. *Landscape and Urban Planning* 61: 181–191.

Tarlow, P. 2007. Tourism: Dealing with the senior market. *Destination World News.* Available at: http://www.destinationworld.info/newsletter/feature36.html. Accessed 20 Nov 2016.

The Economist. 2016. Shades of grey. *The Economist.* Available at: http://www.economist.com/news/britain/21690060-pensioners-incomes-are-now-higher-those-working-households-some-are-doing-much. Accessed 9 Mar 2017.

Tsiotsou, R.H., V. Ratten, and S. Hudson. 2010. Wooing zoomers: Marketing to the mature traveler. *Marketing Intelligence & Planning* 28: 444–461.

Urzi Brancati, C., and B. Franklin. 2015. *Here today, gone tomorrow: How today's retirement choices could affect financial resilience over the long term.* London: International Longevity Centre. Available at: http://www.ilcuk.org.uk/images/uploads/publication-pdfs/Here_today,_gone_tomorrow_1.pdf

Weiss, R.S. 2005. *The experience of retirement.* Ithaca: Cornell University Press.

Rosie Day is a Senior Lecturer in Human Geography at the University of Birmingham, UK. Her research interests centre around social inequalities in access to and experience of environmental and energy resources, with a related interest in environments of ageing and older age. She works collaboratively on a number of multidisciplinary projects in diverse international contexts.

Russell Hitchings is a Senior Lecturer in Human Geography at University College London, UK. His work uses qualitative social research methods to understand the cultural dynamics that shape the ways in which different groups of people come to organise specific aspects of their lives. The broader aim is to find effective ways of harnessing these dynamics in pursuit of healthier and more sustainable societies.

Emmet Fox is a researcher at the Global Sustainability Institute at Anglia Ruskin University, UK. He has researched senior leisure travel for the DEMAND centre while at the University of Birmingham and has published on Pierre Bourdieu and on climate change communication and receptivity in Ireland. He has particular interests in environmental sociology, participatory democracy and social change.

Susan Venn is a Senior Research Fellow at the Centre for the Understanding of Sustainable Prosperity (CUSP) at the University of Surrey, UK, where she is exploring how people in diverse contexts negotiate their aspirations for the good life. Her research interests focus on sustainable lifestyles, everyday practices and life course transitions. Prior to joining CUSP, Sue was a researcher at University College London and was part of the DEMAND Research Centre where she explored older peoples' experiences of and aspirations for future travel beyond retirement.

Julia F. Hibbert is a Postdoctoral Research Fellow in Tourism in the Faculty of Management, Bournemouth University, UK. Her doctoral research explored the role of identity in shaping travel behaviour. She specialises in qualitative research methods and has undertaken work examining aspects of ageing and dementia in tourism. Her research interests include tourism and the sharing economy, sustainable transport systems and disaster and crisis management within tourism.

Part 4

Tracing Trajectories

What would my grandmother say if she could see me now, serving takeaway for our dinners at gone 9 o'clock? Maybe she would be pleased that I wasn't tied to the house like she was. My mum did it all of course, working and cooking a proper meal every day, fresh ingredients. No microwaves or ready meals then, and my Dad never home from work in time to do it, even if he knew how. Sure it's better now, for women at least. Thank goodness for microwaves and washing machines, and what did I do before we had a dishwasher? All the same, I do miss the family meals we had when I was a child, sitting round the table together, every day. But seriously, who these days can get a home cooked meal on the table by six thirty every day, and the family all there to eat it? No way we will ever get back to that being normal.

As made clear in the introductory chapter, time can be understood and worked with in different ways. Whilst the daily and weekly rhythms of everyday life can be revealing of the structuring effects of routine temporal orders, a longer temporal view can reveal different things, especially about trajectories of change. Over months and years, practices evolve, compete, rise and fall in prevalence, circulate among people and through spaces, and structure our time in different and new ways. The three chapters in this part of the book all turn attention to tracing such longer-term change. Whilst their foregrounded questions are quite distinct, the shifting dimensions of working practices figure, both directly and indirectly, in the analytical accounts that they develop. Given that practices inter-relate, 'hang together' in bundles and compete for time, we would expect that trajectories of change in both who is doing work and how work is organised and carried out would have a host of consequences. These include direct and indirect consequences for patterns of energy demand (e.g. related to where and when food is prepared and eaten, the need to travel to meetings with clients and how and when domestic work such as laundry is undertaken). Each of the chapters demonstrates instances of these co-evolving dynamics, along with a range of other insights about longer-term trajectories of change.

Mathieu Durand-Daubin and Ben Anderson's detailed analysis of data from 20+ years of time use surveys in Great Britain and France shows both stability and change in cooking and eating practices. This is detailed in terms of when, where and for how long people engage in these prac-tices, as well as differences in trajectory between the two countries, and between different groups of people. For example, we learn that in both France and Great Britain, eating lunch has decreased in prevalence, espe-cially the practice of eating it at home, with some of this transferred to eating at work. At the same time dinner in both countries, whilst hap-pening later in France than in Great Britain, is being pushed into the evening. Whilst deep insights into how individuals are managing com-peting practices remain necessarily out of reach when working with this type of data, it is surmised that these changes are conditioned by other dynamics, in particular greater employment levels, especially among women, and increasing time pressure on those in employment. Cooking and to some extent eating evidently compete with work for time. Understanding where, when and to what degree energy use is involved in the fundamentally shared, but enormously varied, set of practices that enable eating cannot then be divorced from what else is going on in the ever evolving ordering of society.

Institutional change and the increasing participation of women in the workforce is also a major focus of Mary Greene's chapter, but her method is more attuned to tracing how what she terms the individual's 'practice career' evolves over time, co-incident with the wider context of institu-tional and technological change. Although in the context of Ireland, which has its own specificities and a relatively late trajectory in terms of women entering the workforce in large numbers, the individual biographies of her male and female participants shed some light on how the patterns viewed at an aggregate level in Durand-Daubin and Anderson's chapter are being formed in individual households. Here we see that as women take on employment and other projects outside of the home, the timing of domestic work and its allocation is subject to change. Greene's close study of how this is achieved in households shows us the enabling role of technology, which can reduce the time needed for domestic work as well as allowing temporal flexibility, but also points to an increasing energy intensity of domestic work.

Moving focus from households to engineering consulting firms, Ian Jones and colleagues' contribution also highlights the pivotal role of technology and its affordances in shaping the organisational development of such firms and their working practices. They show how the deregulation of air travel and developments in all kinds of ICT, including design software, have allowed firms to evolve over a period of around 20 years from regionally focused and situated entities to organisations with global reach and dispersed members. Rather than telling the history of these technologies, however, Jones et al. foreground the coordination and cooperation that characterises work within the firms. As ways of procuring work and organising regional offices and professional teams have changed, business travel has been embedded, seemingly intractably, into the firms' operations in order to maintain the physical co-presence of workers and clients, as well as to facilitate flexibility in how expertise is used within the firm for globally dispersed bids. While Greene's contribution then cautions us that employing technology to relieve people of domestic labour and allow them to do other things might lead to increasing energy intensity, Jones et al. see in evolving technology a complex set of potentials. Rather than focusing upon how technologies might try to re-create physical co-presence, they suggest that it would be fruitful to consider how technologies could promote coordinated timespaces of work—facilitating new ways of collaborating, managing projects and responding to colleagues or clients. It is not then virtual working per se that requires further consideration, but rather how corporate (re)organisation shapes working practices with particular temporal and spatial dynamics, prompting related travel and energy demands.

Historical trajectories can thus be traced in different ways, and within different spatial contexts; in Great Britain, France, Ireland and the globally distributed world of business consulting in the chapters in this part of the book. How the spatiality of change is itself evolving is clearly also a pertinent question. Looking ahead, can we still follow the interweaving of practices, technologies, norms and institutional configurations within distinct boundary geographies? Or taking on the arguments made about space in the introductory chapter, are there other geographies now to be worked with? Change is always to some degree situated in place and contextualised by the distinct histories of that place, but it is also often much more than that and subject to forms of shifting relation that have an extended reach. Projecting forward to think about future trajectories of socially embedded energy demand—a crucial and much more involved task than often acknowledged—is then both a question of what is particular to a place, but also what is shared and connected across much broader and more involved geographies of change.

10

Changing Eating Practices in France and Great Britain: Evidence from Time-Use Data and Implications for Direct Energy Demand

Mathieu Durand-Daubin and Ben Anderson

10.1 Introduction

In the global challenge to reduce energy consumption, interventions and injunctions to change behaviour tend to ignore the constantly evolving nature of the practical arrangements to be changed, the fabric of everyday life as it has been, as it is today and how this relates to energy consumption. If many researchers agree that knowing demand is about understanding what energy is finally consumed for, either focusing on consumers or social practices (Janda 2011; Shove and Walker 2014), empirical energy use research still mainly takes into account building and appliance efficiency, environmental constraints and standard socio-economic attributes (age, occupation, employment status). This largely ignores the way that everyday performances of social practices produce

M. Durand-Daubin (✉)
Électricité de France, Palaiseau, France

B. Anderson
University of Southampton, Southampton, UK

© The Author(s) 2018
A. Hui et al. (eds.), *Demanding Energy*, DOI 10.1007/978-3-319-61991-0_10

regular structures based on repetition, synchronisation or sequencing, and so play a significant role in the shaping of demand.

In this chapter, we analyse the historical evolution and current arrangement of practices which are widely shared, highly regular and key markers in the organisation of everyday life: cooking and eating. Our previous work using the most recent time-use surveys in France and Great Britain explored the identification of distinct eating practices (Durand-Daubin and Anderson 2014) and the synchronisation of cooking and peak electricity demand (Durand-Daubin 2016) in an attempt to quantify a part of the large diversity of food practices described in the qualitative literature. In this chapter we take advantage of similar but historical time-use surveys to follow these eating practices over three decades in both countries. The comparative evolution of these practices in France and Great Britain across socio-demographic groups highlights structural change and evolution in the timing, location and synchronisation of eating practices and their consequential energy consumption.

Changes in Social Practices and Energy Demand

This chapter's approach draws on the argument that people use energy as part of accomplishing social practices (Reckwitz 2002; Warde 2005), and it is the variation in the performances of these practices that drives variation in energy consumption. From this it follows that understanding energy demand depends, above all, on understanding the timing, location, context and materiality of a range of inter-connected social practices (Shove 2012; Shove and Walker 2014). In addition, examining the changing temporal distribution of these performances may generate new insights into the way demand for energy has changed and how it may evolve in the future (Shove et al. 2012). That these patterns of activities, and thus their energy demands, change over time is perhaps self-evident (Anderson 2016; Higginson et al. 2013) if apparently rarely seriously considered in energy policy.

Whilst recent calls for improved evidence for energy policy development note that little is known about 'how the use of homes and workplaces by people affects patterns of energy demand' (Department of

Energy and Climate Change 2014: 14), any insights are seen as aids in the surmounting of barriers to 'techno-adoption' rather than to provide suggestions for the re-configuration of social practices. Although the restructuring of domestic practices might appear to be off the public policy agenda in Great Britain, this seems a sizeable missed opportunity. As Higginson et al. emphasise (2013), to assume that practices are inviolable is to claim that they never change. Yet there is substantial empirical evidence that all social practices evolve, albeit at differing rates and with different trajectories (Cheng et al. 2007; Shove 2003; Shove et al. 2012). If practices have changed in the past then they are necessarily open to further change and to intentional intervention (Pullinger et al. 2014).

In the next section we build on the few studies that have engaged with the underlying activities or practices that shape the variation in energy demand to consider the value of time-use data in tracking the recent trajectories of change in cooking and eating practices in Great Britain and France.

Eating Practices and Energy Consumption

A wealth of research has already explored the evolution and diversity of eating practices in France and Great Britain. The majority of studies are rooted in the field of anthropology. For example, Mennell (1996) covers the social history of eating practices from the Middle Ages to the 1980s, in England and France, in his extensive *All Manners of Food*. In France, Kaufmann focuses on the social functions of meals (Kaufmann 2005, 2007), while Garabuau-Moussaoui et al. (2002) use the lens of emerging practices among younger generations to describe the diversity of eating practices across the world and their current transformation.

Quantitative research analysing the evolution and diversity of the time and money spent in these eating practices has been based on national economic statistics and time-use surveys, and reflects key policy foci ranging from hunger and obesity issues to the division of paid and informal work. Warde and Martens (2000) produced an extensive analysis of "eating out" practices in three urban areas based on a mixture of qualitative interviews, ad hoc quantitative survey and public statistics (Warde and Martens 2000). Southerton et al. (2012) give a glimpse of the diversity in the timing of eating practices in a cross-cultural comparative analysis of Spain and Great Britain, based on time-use surveys (Southerton et al. 2012) although this analysis confines itself to the duration rather than the timing of eating. Beyond meal structure and duration, Yates and Warde (2015) also enter into the evolution of the content of meals in Great Britain based on two ad hoc surveys, finding more variations in the food eaten than in the meal pattern (Yates and Warde 2015). In France, recent evolutions in time use were analysed by De Saint Pol (2006), showing the persistence of the three meals a day model (De Saint Pol 2006) and the development of the sociability function of meals across different social categories (Larmet 2002). In addition, national statistics on food expenditure show the decline of meat and raw products compared to ready meals and processed food and the increasing cost of eating out (Larochette and Fernandez-Gonzalez 2015).

In contrast to this focus on food consumption per se, this chapter is an attempt to describe the link between these prior social understandings of eating practice evolutions and their specific implications for household energy consumption. Eating requires a number of steps which consume energy, from the production of food to the disposal of waste, including cooking and eating meals. Each of these steps can be handled in many different ways, involving various material arrangements (land, infrastructures, chemicals, building, appliances, fuels) and consumption levels. The content of the meal has an upstream impact on the energy embedded in food, on the energy and appliances final consumers will need to store and prepare food. Here we focus on the everyday aspect, the final part of this chain that is the tip of the 'energyberg' that consumers are involved in: eating and cooking. Although these steps may not consume the largest part of the energy required for the entire process (Clear et al. 2013), they constitute the justification or the final objective of the preceding energy consuming processes.

Recent qualitative research has explored the link between these practices and energy consumption, from cooking styles (Clear et al. 2013) to the coordination of cooking between household members (Isaksson and Ellegaard 2015). However this chapter focuses on changes in eating and cooking times, synchronisation, places and frequencies because they directly and indirectly affect the time when energy is consumed. As a result they play a major role in setting the rhythm of everyday life (Southerton et al. 2012), defining meaningful periods of time, synchronising practices and thus conditioning the rise of evening peak electricity demand in interaction with natural lighting (Durand-Daubin 2016) and other evening weekday activities (Torriti et al. 2015). The place where the food is consumed has implications not only for the space heated and lit during the meal, but also on where and how the cooking is done. Cooking is an energy intensive activity, especially when it involves heating processes, and so the variations in the frequency and duration of the food preparation reported by consumers provide indications of the amount of energy consumed at home for this provision.

Drawing on these multiple links with energy, the chapter presents analysis of trends in the frequency, timing, location and social distribution of eating and cooking activities during the week, and especially in the evening when everyday practices are highly constrained and electricity demand at its highest. In so doing, the chapter will highlight the evolving configuration of energy demanding practices across a range of social dimensions.

10.2   Analysis of Changes in Eating and Cooking Practices

Data and Methods

The analysis presented in this chapter uses two comparable historical time-use survey resources. For Great Britain, this is the recently developed Multinational Time Use Study (Gershuny et al. 2012) which includes detailed activity sequences with 'harmonised' activity codes from representative British time-use diary surveys carried out in 1974, 1983, 1987, 2000 and 2005 (Table 10.1). The French data come from the 1985–1986, 1998–1999 and 2009–2010 (Table 10.2) surveys carried out by the Institut National de la Statistique et des Etudes Economiques (Brousse 2015). As the tables make clear, comparative analysis over time must take into account changes in coding schemes, data collection methods, sampling and response details (Anable et al. 2014; Anderson 2016).

Table 10.1 Summary of UK time-use data

Year	Sample	Size & season	Time interval	Format	Eating	Cooking & food preparation
1974	All 5+ in representative household sample	2598 February, March, August, September	30 minutes	7 diary days, primary & secondary activities (73 codes), location known, co-presence unknown	34 Eat meals, snacks 35 Meal break, dinner break, at home 38 Drink tea, coffee etc., at home 39 Drink alcohol, at home 49 Meal break at work	53 Prepare meals or snacks
1983	Representative sample 14+	1350 January, February, September, November, December	15 minutes	7 diary days, primary & secondary activities (188 codes), location known, co-presence of others known	0103 Scheduled break at work (meal) if time = 15 minutes 1502 Drinking non-alcoholic beverages + location in workplace or school 0402 Lunch break at educational establishment – school 1501 Eating at home 1502 Drinking non-alcoholic beverages	0601 Food preparation 0602 Bake, freeze foods, make jams, pickles, preserves, dry herbs 0604 Make a cup of tea, coffee
1987	Representative sample 14+	1586 March – June	15 minutes	7 diary days, primary & secondary activities (190 codes), location known, co-presence of others known	As for 1983	As for 1983

(continued)

Table 10.1 (continued)

Year	Sample	Size & season	Time interval	Format	Eating	Cooking & food preparation
1995	Representative sample 16+	1962 May	15 minutes	1 diary day, primary activities only (31 codes), location & co-presence of others unknown	4 Eating/home	3 Cooking
2001	All 8+ in representative household sample	8688 All months	10 minutes	7 diary days (weekday & weekend), primary & secondary activities (265 codes), location known, co-presence of others known	210 Eating (at work or school – where = 4) 1310 Lunch break 210 Eating not coded in categories 5 or 38	3100 Unspecified food management 3110 Food preparation 3120 Baking 3140 Preserving
2005	Representative sample 16+	4854 March, June, September, November	10 minutes	1 diary day, primary & secondary activities (30 codes), location known, co-presence of others unknown	Pact = 4 (eating/ drinking) and pact = 1 (home) or missing	Pact = 5 (preparing food)

Table 10.2 Summary of French time-use data

Year	Sample	Size & season	Time interval	Format	Eating	Cooking & food preparation
1966	Random sample of 18–65 in 7 medium cities + Paris only	4840 February – March 1966 & 1967	5 minutes	1 diary day		
1974	Random sample 18+ in urban areas only	6640 May 1974 – April 1975	5 minutes	1 diary day		
1985	Random population sample 15+ and partner	10,373 households (16,047 individuals) September 1985 – September 1986	5 minutes	2 diary days, main & secondary activity, location (home, work/school, friends/family), mode of transport, presence of others	141 Small breakfast 142 Breakfast 143 Dinner at home 144 Collation, sandwich at home 145 Aperitif at home 146 To have a tea or coffee at home 147 To have a snack at home 154 Collation sandwich away from home 155 To have an aperitif away from home Location! = home 152 Meal at restaurant, snack bar, fast-food 156 Tea, coffee, patisseries in café, bar, tea room 157 Queuing at the canteen, restaurant, café	311 Meals preparation and cooking 312 Peeling of fruits and vegetables

(continued)

Table 10.2 (continued)

Year	Sample	Size & season	Time interval	Format	Eating	Cooking & food preparation
1998	Random population sample 15+ and partner	8186 households (15,441 individuals) February 1998 – February 1999	10 minutes	2 diary days, main & secondary activities, location (home, work/school), mode of transport	141 Meals at home alone or with household members 143 Meals away from home alone or with people of the household and location is another home 144 Meals at home with people not from the household 146 Meals away from home with people not from the household and location is another home	310 Cooking with an associative aim 311 Preparation and cooking of meals 314 Bake conserves, cakes and confectionary 319 Cooking for another household
2010	Random population sample 15+ and partner	10,675 households (16,242 individuals) September 2009 – September 2010	10 minutes	2 diary days; main & secondary activity, location (home, work/school, friends/family), mode of transport, presence of others	141 Meals at home alone or with household members 143 Meals away from home alone or with people of the household and location is another home 144 Meals at home with people not from the household 146 Meals away from home with people not from the household and location is another home	311 Preparation and cooking of meals 314 Bake conserves, cakes and confectionary 319 Cooking for another household

In the case of Great Britain, earlier diaries used longer (30 or 15 minute) activity recording slots whilst later surveys used 10 minutes so that direct comparison of the reported number of episodes of a particular activity is not a robust method of analysing change over time. Furthermore the diaries of 1983 and 1987 were only recorded in specific months (see. Table 10.1) and so are usually pooled to form a full year '1985' (Gershuny et al. 2012) whilst the diaries for 1995 were only completed in May and so are rarely used. However, as Table 10.1 shows, recorded time-use codes for food preparation and eating remained relatively unchanged thus providing the opportunity to analyse change over a 30-year period from 1974 to 2005.

Although time-use surveys were carried out in France prior to 1985 (see Table 10.2) these did not use representative population samples before 1985 and so are excluded from this analysis. As with the British data, the relatively constant coding of food preparation and eating (see Table 10.2) and the similar study designs in the three last surveys allow comparison not only between countries but also between different subgroups of individuals based on socio-demographics.

As Tables 10.1 and 10.2 show, each of the activities reported was also located in varying degrees of detail although these location categories varied from one survey to another. Thus, whilst it would have been preferable to be able to locate 'eating at home' vs. 'eating out' vs. 'eating at friends/family', this was not possible in all surveys.

Finally the lack of a more recent UK national time-use survey (Fisher and Gershuny 2013) forced the use of Great Britain 1974–2005 surveys alongside the French 1985–2010 surveys. Nevertheless, the ability to construct a 20- to 30-year history of the timing, sequencing and, in some surveys, the detailed location of food-related activities provides a substantial basis for empirical analysis of changing practices over time.

As with recent studies of domestic laundry and water demand more generally (Anderson 2016; Browne et al. 2014), while we are unable to explore the detailed contingencies or interconnected networks of materiality and meaning that are part and parcel of the moment by moment performance of practices (Warde 2005), we are nevertheless able to describe and analyse the variation in the temporal structure of the activities that represent the footprints of these performances. This in turn will allow us to draw some conclusions about the changing patterns of their consequential energy demand.

Analytic Approach

Although all of the time-use surveys used capture both primary and secondary activities, the latter were discarded because of the potential influence of variations in the instrument design on their reporting rates (Anderson 2016). In addition, to maintain comparative populations, only individuals aged 16 or over were selected for analysis. Furthermore, to counteract the problem of differing levels of detail in the collection of activity durations, the occurrence of the activity of interest at least once in each half hour interval (30-minute time slot) was used as a comparable indicator across years (Anderson 2016). Although imperfect, as it is unable to represent the duration of activities, this approach not only enables meaningful comparisons across time but also has the benefit of mapping on to the familiar half hour electricity system 'settlement periods' (Darby 2010). Levels of eating and cooking activities were then analysed in two ways: daily dynamics and number of half hours during which the activity was reported for specific meals, places or populations.

Daily dynamics are represented by the percentages of the population reporting the activity in each half hour of the day and are used to study the timing and synchronisation of eating and cooking practices.

The number of half hours in which the activity occurred for one individual in one day is a proxy for the extent of the activity in time. It is used in aggregate form (mean number of hours per social group) to analyse the importance of types of meal and eating locations, in different years and socio-demographic categories.

Drawing on our interest in the relationship between cooking, eating and electricity demand, we focus on the evolution of two meals, lunch and dinner, defined based on the time of eating activities. Our interest for lunch comes from the possibility to have this meal at home or at the workplace or school, allowing the observation of transfers between places where eating and cooking are provided and require energy. Dinner on the other hand is likely to interact with evening peak of electricity demand.

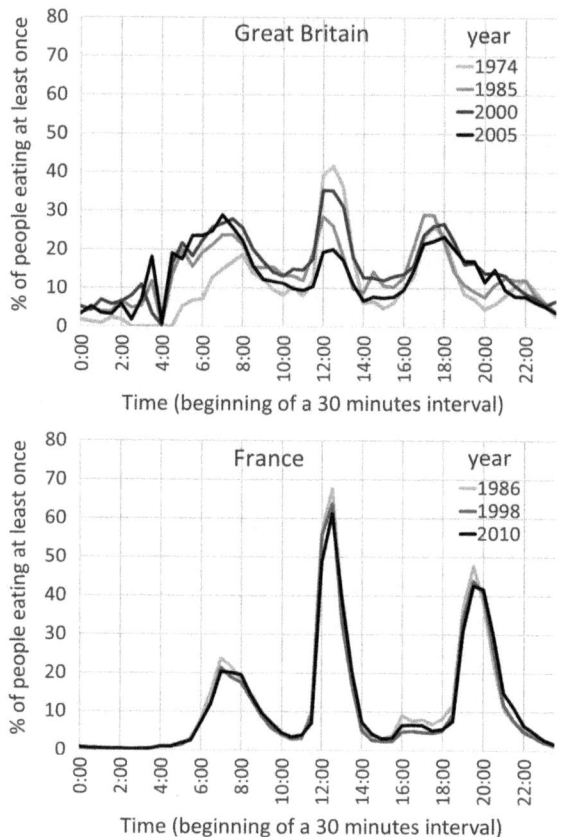

Fig. 10.1 Distribution of eating across time of day by year of survey

Drawing on Figs. 10.1 and 10.2 below, these meals were coded according to the boundaries described in Table 10.3 which allow for the differing timings of eating in Great Britain and France that we describe below.

Lunch: Diversity and Evolutions of Time, Preparation and Place

Lunch is more synchronised in France than in Great Britain. At noon, 60 % of the French population report eating in the 30 minutes from 12:30 to 13:00 (Fig. 10.1), when the British peak only reaches 20%. Cooking

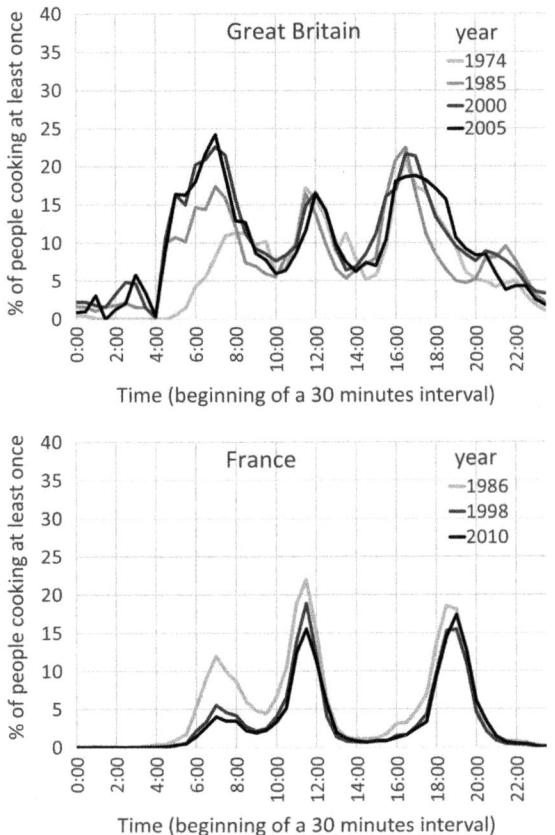

Fig. 10.2 Distribution of cooking across time of day by year of survey

Table 10.3 Coding of lunch and dinner

| | Eating | | Cooking | |
	Start	End	Start	End
Weekday lunch	11:30	14:00	10:00	12:30
Weekday dinner	18:30	22:00	16:00	21:00

is also more concentrated and happens 30 minutes earlier in France (11:30 vs. 12:00) (Fig. 10.2). However the percentage of people involved in cooking at the same time is about the same in both countries (15%), showing a much higher rate of food preparation relative to the number of people eating lunch in Great Britain.

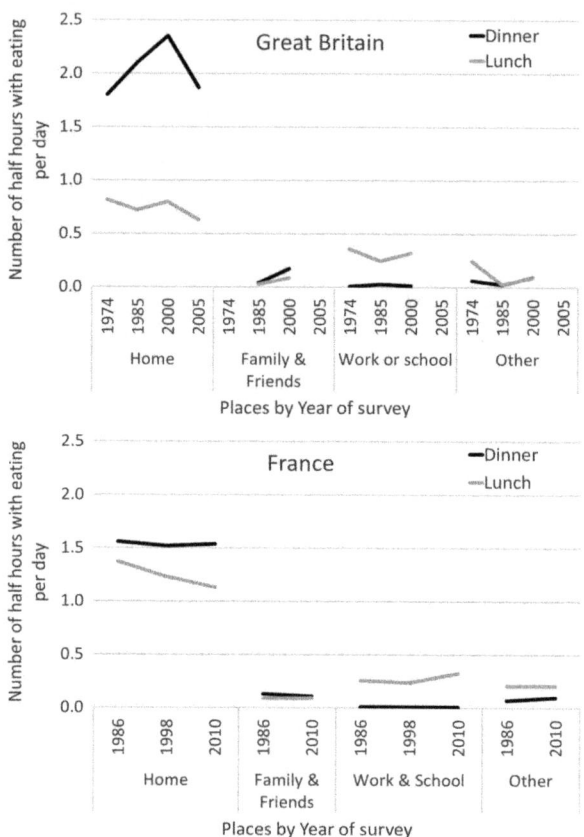

Fig. 10.3 Location of eating meal types

Less food preparation intensity can reveal several phenomena: externalisation (lunch was prepared out of the household) possibly in relation to eating out, mutualisation (one person cooking for more people), or simpler content (food can be eaten with less preparation). The place of eating (Fig. 10.3) reveals that even if in France lunches are more often eaten in other places (e.g. restaurant) than in Great Britain, they still happen more frequently at home. Hence, there is slightly more externalisation in France, but there is also probably more shared lunches at home. Eating at work happens as often in

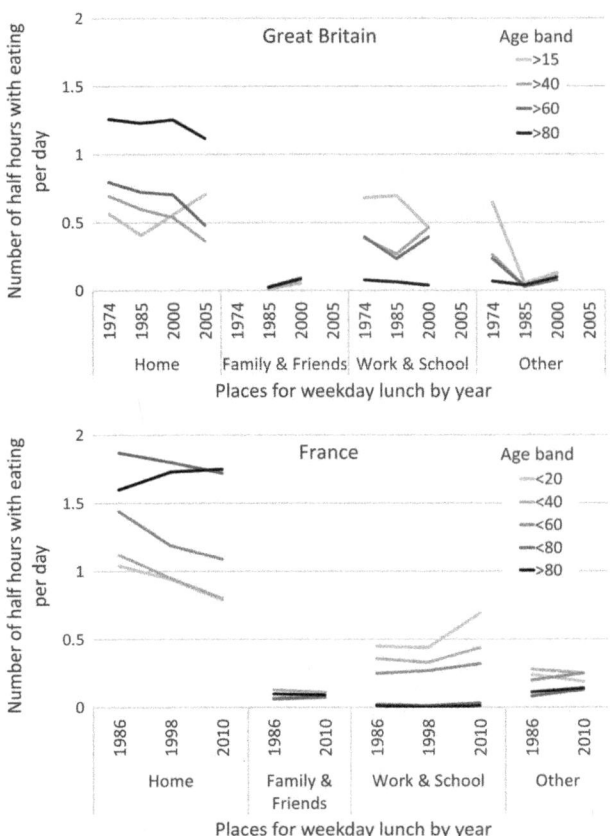

Fig. 10.4 Weekday lunch by age and place

both countries and is related to people's employment status, and consequently age (Fig. 10.4): students (more frequent in lower age bands) eat out more often than those in work, who in turn eat out more often than people after the age of retirement. Students and those in work also eat more in other public places, especially in France. Partly resulting from the demographic structure of eating out for lunch, cooking decreases with the level of household income in both countries (Fig. 10.5). Regarding cooking mutualisation, in both countries, women cook lunch more often, but the difference is

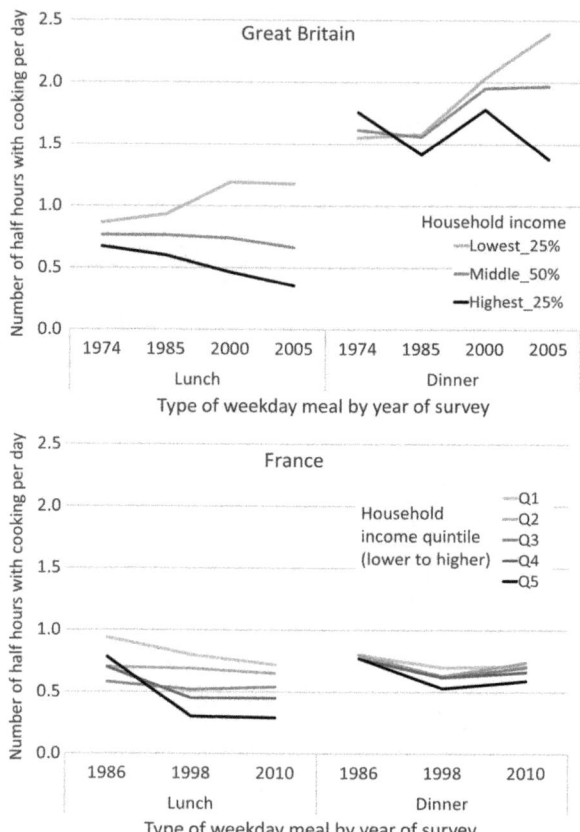

Fig. 10.5 Patterns of cooking/preparing meal types by income

relatively larger in France (+350 % vs. +50 % more than men), which could contribute to more lunches at home with the same amount of cooking in this country.

The comparison of the timing and synchronisation of these activities from year to year also reveals differentiated evolutions between the countries. In both cases the time that lunch is eaten has remained largely unchanged but eating lunch has decreased (Fig. 10.1), but in much larger proportion in Great Britain than in France (−20 % vs. −7 % of participation at peak time). In Great Britain a part of the decrease in eating lunch appears to be compensated by an equivalent increase in eating breakfast, especially at earlier hours (between 5:00 and 7:00).

In both countries, the decrease is steeper for lunches at home, while eating at work increased slightly in the most recent periods (Fig. 10.3), potentially indicating a small transfer from home to the workplace. This transfer is clearer for younger groups, students and working people (Fig. 10.4), except for the youngest population in Great Britain who experienced the exact opposite redistribution from the workplace and school to home. The general shift of lunch from home to the workplace is mainly explained by the increase in the population rates of employment and students, especially among women, in both countries, during this period.

In France, among employed people, the practice of going back home for lunch also decreased. However, this change was not compensated by more lunch in the workplace in this population.

Eating less at home correlates with a decrease in lunchtime cooking in France but not in Great Britain (Fig. 10.2). The sustained level of midday cooking when the level of eating lunch at home declined may result from a simple contraction of the time spent in eating for a constant number of lunches, or less sharing of cooking, due to only a part of the household not eating lunch at home, or an increase in the type of preparation needed due to a change in content. In France cooking lunch decreases in every age band while in Great Britain, it is mainly seen for active people between 20 and 60, while younger people cook much more than they used to.

The difference in the rates of cooking and eating lunch between higher and lower income groups has widened during the period studied. In France, cooking among people with the highest income declined quickly and substantially, before stabilising in the last period. The rate of cooking in the lower quintile declined later and more slowly but continues to decrease. In contrast, in Great Britain the level of cooking in the lowest quartile increased although, as in France, it constantly decreased in the highest quartile. The apparent stability of lunch time cooking in Great Britain may therefore be the result of opposite social trends: active working people with high incomes may cook less, while younger and older poorer people cook more.

Dinner: Diversity and Evolutions of Time, Preparation and Place

Currently, dinner tends to happen earlier in Great Britain, at around 18:00, but two hours later in France, at 20:00 (Fig. 10.1), where it comes after a long period of four and a half hours with barely any eating activity. Before that a very small frequency of tea break and "goûter" between 4:00 and 5:00 can be observed in both countries. Cooking happens one hour before dinner at 17:00 and 19:00. As with lunch, dinner is more synchronised in France than in Great Britain (42 % vs. 23 % of participants at peak time). In Great Britain cooking and eating dinner extend over very long periods of time (see Fig. 10.2) with a higher proportion of British respondents involved in cooking (over 15 % in each half hour from 16:00 to 18:30) than is the case in France (only greater than 10 % between 18:00 and 19:30).

Over time, dinner has come to be eaten and cooked later in the evening especially in Great Britain with substantial change having taken place between 1974 and 1985, before a period of slower rate of change, confirming Cheng et al.'s results for the reduction in total time spent preparing food (Cheng et al. 2007: 46). The analysis also highlights the disappearance of late evening food preparation which may have been associated with preparing food for the next day or for a late 'supper' (Cheng et al. 2007).

As Fig. 10.3 shows, dinner is overwhelmingly eaten at home in both countries throughout this period, with an unexplained peak in Great Britain in 2000. Eating dinner at family or friends' increased slightly in Great Britain (but not France), whilst eating dinner at restaurants (other places) increased only in France. Younger people and single households eat out and at other peoples' homes more often and the level of eating at friends' by income quintiles has tended to converge over time. The effect of employment and income on the eating of dinner is much lower than on lunch, although students report eating less often at home, and more at friends or family. In France, while the amount of eating and cooking didn't change much for dinner, people in the same household tend to share their dinner together less often than they used to. Conversely, meals shared with people not belonging to the household have increased, in particular for the lower income groups who appear to be catching up with wealthier people who were receiving guests more often in the past (Larmet 2002). On the other hand, these trends do not give a clear direction for changes in the level of shared eating and cooking which are also significant for understanding the energy intensity of dinner.

As already observed for lunch, women cook dinner more often than men. However, as we can see from Fig. 10.6, in both countries, dinner is the meal for which men's involvement increased the most, especially in Great Britain, even if it remains far below the level of women cooking, which stays stable in Great Britain and declines in France. Further analysis by household type suggests that due to sharing, larger households also cook less per person than single ones and the amount of cooking per person decreased more for single parents than for the other categories.

10.3   Discussion

Despite the difficulty of analysing varying units and categories of measurement over time, the analysis has suggested that there are observable changes in the level, timing, and location of eating and cooking that will have had an impact on the patterning of energy demand associated with eating practices.

Evolution in the Structure and Timing of Meals

As previously described (De Saint Pol 2006; De Saint and Ricroch 2012), the three meals structure is more stable and synchronised in France than in Great Britain, where eating and cooking spread more and more across the day. Common changes can be observed in both countries, even if they are slower in France. During the week, eating and cooking activities have been reduced, anticipated, delayed or shared because of the increasing

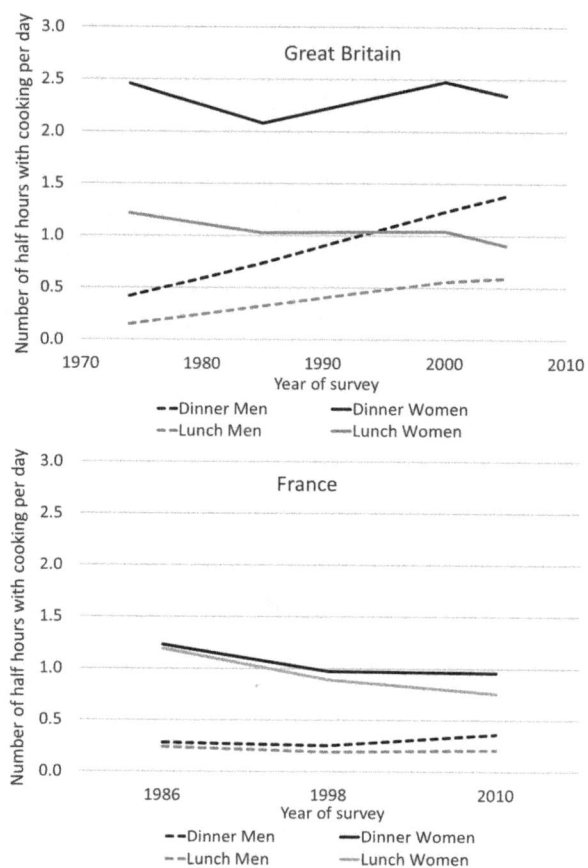

Fig. 10.6 Gender of cooking/preparing meal types

time pressure received from work, in a similar way to what has been observed for laundry (Anderson 2016). From these changes we can expect a consequent decrease in energy demand for eating lunch at home and an increase at the beginning and at the end of the day, in Great Britain. These changes are likely to have altered peak energy demand in several ways. On the one hand spreading demand overall reduces peaks. On the other hand, eating later in the evening contributes to higher peak electric demand because of a larger overlap between cooking and other evening activities (Durand-Daubin 2016).

However, the impact of these changes in eating rhythms on energy demand is very indirect. More time spent, or more frequent eating episodes, does not necessarily mean more energy consumption, but their distribution and movement in time and space can help to understand changes in the pressure put on specific moments and places. Energy demand for eating will also vary depending on how it translates into cooking.

Cooking at Home

In both countries, the vast majority of the eating events happen at home, allowing for a variety of domestic food preparation and different levels of energy consumption, among which only reported (or absent) cooking events can be differentiated from time-use surveys. In this case, the food preparation associated with each eating event can vary and have an impact on the associated direct energy consumption. The type of food preparation depends on the place of eating but also on possible externalisation of the cooking alone: deliveries from restaurants or ready cooked meals from the industry. The energy implications of a reported episode of 'food preparation' are therefore currently extremely unclear.

Overall cooking declined in France but not in Great Britain where opposite trends appear to cancel each other. Changes in the amount of eating at home generally translate into associated change in the amount of cooking: decrease in French lunches, increase in British dinners. However this is not the case for the decrease in British lunches at home where we see the same level of cooking. The explanation for this apparent discrepancy is not known, but could simply result from a contraction of the duration of cooking for a given number of lunches.

The evolution of cooking is also highly differentiated by the level of income with higher income groups consistently and increasingly reporting less cooking. Employment status and the ability to eat out only account for a small part of this difference, implying the impact of quicker cooking practices or other developments such as takeaways or deliveries, which are more difficult to detect in time-diary data.

Finally, the effect of this decline in the time spent on cooking at home in France does not appear clearly in the evolution of the energy consumption spent in residential cooking (ADEME 2012). Estimated cooking consumption per household has decreased, but the amount of energy consumed per person has remained stable, given the reduction in the average size of households.

Outsourcing Energy

Variations in the place for eating strongly depend on the type of meal and surrounding activities. Weekday lunch is the best opportunity to eat at work, school or restaurants, with those in work and in higher income groups more likely to report doing this in each country. If the number of lunches taken at work increased, it is partly because the number of people employed increased, which was driven by women employment across the period studied, not because those in work eat more often at the workplace. In France, the share of economically active people going back home for lunch decreased, but this meal at home appeared to be skipped rather than replaced by a meal elsewhere.

Eating out at restaurants increased only for wealthier people in France for the meals studied, while in Great Britain income differences in this practice have narrowed. More specifically, in France, eating out declined for lower incomes and unemployed people, and increased for retired and wealthier people, thus most likely increasing home-based energy consumption amongst those least able to afford it although potentially reducing expenditure on eating out.

10.4   Conclusions and Future Directions

This study of eating schedules on weekdays across time, socio-demographic groups and two countries has revealed both changing and stable aspects in the practices associated with preparing and eating lunch and dinner, as well as elements of the social transformations and constraints that disrupt or reinforce them. Domestic energy demand arising from eating and cooking may have been reduced at midday and pushed later in the evening, partly as a result of increasing time pressures and employment rates, especially among women. The reduction in midday energy demand, when there is often surplus generation capacity, and the potential additional and extended loading of evening energy demand have clear implications for both countries, where evening peak demand is generally met through costly and high carbon generation.

As a partial counter-balance, food consumption has also expanded outside homes to the services and industrial sectors, implying more anticipation and spread across the day, and differing material arrangements. This in turn may lead to increased economies of 'energy scale' coupled with reductions in domestic energy consumption, but only for certain groups in society.

Consequently, interventions aiming at saving energy or reducing peak demand will need to fit this frame of change and constraint. Adding financial or normative incentives to avoid consumption in the evening may increase the constraints on this meal and accelerate existing trends toward later meals and, marginally, eating out for wealthier people. However this acceleration is likely to remain limited in France given the apparent high resistance of the intensity and scheduling of this practice to the impact of much stronger social changes in the past. Effective interventions would therefore need to engage with practices of time management and the stickiness of evolving domestic/paid labour arrangements.

In conclusion, whilst the analysis has produced insights of potential value, we acknowledge that the standardisation of the categories which is necessary for quantitative analysis and the conversion of durations to binary indicators of activities within a half hour reduces the accuracy of the analysis. It may also miss meaningful differences in the materiality and purpose of the practices that emphasise stable aspects of the practices. More complex analyses aggregating eating episodes into proper meals would allow deeper analysis of meal durations, participations and direct link with cooking. A distinction between the number of people taking part in a practice and the intensity of the practice by practitioners would allow the production of more detailed analysis of the evolution of eating. Considering the sequences of activities in which meals happen would also help to better describe the constraints on these practices. A more complete analysis would also include consideration of other types of meal such as breakfast and weekend meals, and more detailed analysis by socio-demographic groups to develop deeper insights into not only what has changed, but also the potential future direction of change.

Finally, time-use diaries would greatly benefit from triangulation with external data to complement our understanding of changes and to strengthen the link between people's activities and energy consumption. These external data could cover evolutions and diversity in several fields: content of meals based on food consumption and expenditures, features of domestic appliances, available food and catering supply and associated processes and energy consumption levels. Taken together these data would support substantially more detailed analyses of the changing practices of eating and their implications for current and future energy demand.

Acknowledgements This work was supported by the Engineering and Physical Sciences Research Council [grant number EP/K011723/1] as part of the RCUK Energy Programme and by EDF as part of the R&D ECLEER Programme.

Bibliography

Agence de l'environnement et de la maîtrise de l'énergie. 2012. *Energie et climat – les chiffres cle 2012*. Angers: Agence de l'environnement et de la maîtrise de l'énergie. Available at   http://multimedia.ademe.fr/catalogues/chiffres-cles-energie-climat-2012/data/catalogue.pdf

Anable, J., B. Anderson, E. Shove, et al. 2014. *Categories, concepts and units: Energy in and through time*. Lancaster: DEMAND Centre, University of Lancaster. Available at http://www.demand.ac.uk/wp-content/uploads/2014/07/Working-Paper-3.pdf. Accessed 25 Feb 2015.

Anderson, B. 2016. Laundry, energy and time: Insights from 20 years of time-use diary data in the united kingdom. *Energy Research and Social Science* 22: 125–136.

Brousse, C. 2015. La vie quotidienne en France depuis 1974. Les enseignements de l'enquête emploi du temps. *Economie et Statistique* 478: 79–117.

Browne, A.L., M. Pullinger, W. Medd, et al. 2014. Patterns of practice: A reflection on the development of quantitative/mixed methodologies capturing everyday life related to water consumption in the UK. *International Journal of Social Research Methodology* 17: 27–43.

Cheng, S.-L., W. Olsen, D. Southerton, et al. 2007. The changing practice of eating: Evidence from UK time diaries, 1975 and 2001. *The British Journal of Sociology* 58: 39–61.

Clear, A.K., M. Hazas, J. Morley, et al. 2013. Domestic food and sustainable design: A study of university student cooking and its impacts. *Proceedings of the SIGCHI Conference on Human Factors in Computing Systems (CHI '13)* 2013: 2447–2456.

Darby, S. 2010. Smart metering: What potential for householder engagement? *Building Research & Information* 38: 442–457.

De Saint Pol, T. 2006. Le dîner des français: Étude séquentielle d'un emploi du temps. *Economie et Statistique* 400: 45–69.

De Saint Pol, T., and L. Ricroch. 2012. Le temps de l'alimentation en France. *Insee Première* 1417: 1–4.

Department of Energy and Climate Change. 2014. *Developing DECC's evidence base*. London: Department of Energy and Climate Change.

Durand-Daubin, M. 2016. *Cooking in the night: Peak electricity demand and people's activity in France and Great Britain*. DEMAND centre conference 2016, Lancaster.

Durand-Daubin, M., and B. Anderson. 2014. *Practice hunting: Time use surveys for a quantification of practices distributions and evolutions*. BEHAVE 2014 – Paradigm shift: From energy efficiency to energy reduction through social change, Oxford.

Fisher, K., and J. Gershuny. 2013. The 2014–2015 United Kingdom time use survey. *Electronic International Journal of Time Use Research* 10: 91–111.

Garabuau-Moussaoui, I., D. Desjeux, and E. Palomares. 2002. *Alimentations contemporaines*. Paris: L'Harmattan.

Gershuny, J., K. Fisher, E. Altintas, et al. 2012. *Multinational time use study, versions world 5.5.3, 5.80 and 6.0 (released October 2012)*. Oxford: Centre for Time Use Research.

Higginson, S., M. Thomson, and T. Bhamra. 2013. "For the times they are a-changin": The impact of shifting energy-use practices in time and space. *Local Environment* 19: 520–538

Isaksson, C., and K. Ellegaard. 2015. Dividing or sharing? A time-geographical examination of eating, labour, and energy consumption in Sweden. *Energy Research & Social Science* 10: 180–191.

Janda, K.B. 2011. Buildings don't use energy: People do. *Architectural Science Review* 54: 15–22.

Kaufmann, J.-C. 2005. *Casseroles, amour et crises. Ce que cuisiner veut dire.* Paris: Armand Collin.

———. 2007. *Familles à table. Sous le regard de Jean-Claude Kaufmann.* Paris: Armand Collin.

Larmet, G. 2002. La sociabilité alimentaire s' accroît. *Economie et Statistique* 352: 191–211.

Larochette, B., and J. Fernandez-Gonzalez. 2015. Cinquante ans de consommations alimentaires, une croissance modérée mais de profondes transformations. *Insee Première* 1568: 1–4.

Mennell, S. 1996. *All manners of food: Eating and taste in England and France from the middle ages to the present.* Chicago: University of Illinois Press.

Pullinger, M., B. Anderson, A. Browne, et al. 2014. New directions in understanding household water demand: A practices perspective. *Journal of Water Supply: Research and Technology – AQUA* 62: 496–506.

Reckwitz, A. 2002. Toward a theory of social practices. *European Journal of Social Theory* 5: 243–263.

Shove, E. 2003. *Comfort, cleanliness and convenience: The social organisation of normality.* London: Berg.

———. 2012. Putting practice into policy: Reconfiguring questions of consumption and climate change. *Contemporary Social Science* 9: 415–429.

Shove, E., and G. Walker. 2014. What is energy for? Social practice and energy demand. *Theory, Culture & Society* 31: 41–58.

Shove, E., M. Pantzar, and M. Watson. 2012. *The dynamics of social practice: Everyday life and how it changes.* London: Sage.

Southerton, D., C. Díaz-Méndez, and A. Warde. 2012. Behavioural change and the temporal ordering of eating practices: A UK-Spain comparison. *International Journal of Sociology of Agriculture & Food* 19: 19–36.

Torriti, J., R. Hanna, B. Anderson, et al. 2015. Peak residential electricity demand and social practices: Deriving flexibility and greenhouse gas intensities from time use and locational data. *Indoor & Built Environment* 24: 891–912.

Warde, A. 2005. Consumption and theories of practice. *Journal of Consumer Culture* 5: 131–153.

Warde, A., and L. Martens. 2000. *Eating out: Social differentiation, consumption and pleasure.* Cambridge: Cambridge University Press.

Yates, L., and A. Warde. 2015. The evolving content of meals in Great Britain. Results of a survey in 2012 in comparison with the 1950s. *Appetite* 84: 299–308.

Mathieu Durand-Daubin is a Researcher in the Social Science team at the Électricité de France's Research and Development Division. He has a degree in applied statistics. His research focuses on modelling the diversity and evolutions of households' energy consumption, and relies on mixing complementary methods (qualitative and quantitative surveys, monitoring).

Ben Anderson is a Senior Research Fellow in the University of Southampton's Energy and Climate Change Division. He has a degree in Biology and Computer Science and a PhD in Computer Studies. His career has spanned commercial and academic research, focusing on what people do with technologies and how this impacts service and resource provision infrastructures. Whilst his early work focused on understanding the use of media and communications technologies through the triangulation of survey, time-use and usage logging data, he now applies similar approaches to the study of energy demand, working with academic, policy and commercial partners.

11

Paths, Projects and Careers of Domestic Practice: Exploring Dynamics of Demand over Biographical Time

Mary Greene

11.1 Introduction

Understanding the patterning of energy demand at different scales and temporalities is essential for comprehending dynamics in everyday consumption practices. An important but under researched temporal scale of analysis for energy demand is that of biographical time. In Ireland and beyond, momentous change in socio-cultural and techno-material landscapes have radically transformed the way everyday life is experienced and performed over the lifecourse, with major implications for how energy is demanded in the household. However, to date our understanding of how and why patterns of domestic energy demand change over biographical time within the context of wider changes in society remains poor.

Recent research has shown how consumption of energy is an outcome of peoples' participation in social practices, such as eating, getting to work or fulfilling social roles, such as parenting (Warde 2005). In foregrounding practices as the key unit of analysis, existing work has

M. Greene (✉)
NUI Galway, Galway, Ireland

demonstrated that practices develop careers or biographies of their own, with their lives often predating and recursively co-evolving with individuals' performances (Shove et al. 2012). To date, the predominant practice-theoretical approach to exploring dynamics over time has been to explore the biographies of practices themselves with the consequence that there has been little exploration of individuals' lives. This is despite the importance of individuals' biographic experience for shaping patterns of performance, reproduction and change over time (Shove and Pantzar 2007; Warde 2005). Individuals, too, have a history and future of performances that are intertwined within a life. Shove et al. (2012: 39) stress the importance of individual and collective "careers" of engagement with practice for patterning of energy demand, reminding us that "the lives of practitioners and practices interact" in the dynamics of practice as it plays out over time and space. However, a paucity of research has explored interactions between the lives of individuals and practices in a systematic way.

In response to calls for greater attention to the role of individuals in practice dynamics, a small but growing body of biographically situated practice-theoretical research on demand has emerged. Moving beyond individualised-rationalist lifecourse approaches to consumption (e.g. Lanzendorf 2010) this work is broadening the scope of analysis to advance highly contextualised and dynamic approaches to energy demand (cf. Greene and Rau 2016). Studies that focus on particular life transitions can be distinguished from work that adopts longer temporal frames to consider dynamics in the context of the whole lifecourse. This distinction is important as the temporal scale chosen has implications for studying and understanding mechanisms and processes of reproduction and change.

Adopting a shorter temporal frame, life-transition centred practice research has directed attention to changing logics and contexts of practice during biographic transitions as well as how these intersect with pre-structured and dynamic social identity roles (Burningham et al. 2014; Jaeger-Erben and Offenberger 2014). However, approaches that limit the temporal frame of analysis to focus on dynamics as they are occurring during specific lifecourse transitions are less suitable for exploring processes and mechanism operating over longer time scales.

Work adopting lengthier temporal perspectives point to the value of longitudinal approaches for exploring broader contextual processes shaping patterns of social reproduction and change. Reconstructive-biographic-practice research investigating dynamics in the context of the whole lifecourse is directing attention to the role of socio-historical time and space in shaping or foreclosing the development of practice careers and patterning of energy demand over individuals' lives (cf. Greene and Rau 2016; Hards 2011; Henwood et al. 2015). These studies highlight the dynamic, recursive interaction between socio-technical transformations and the structure and allocation of practices in daily life.

Despite these advances, questions remain regarding how wider structural processes, including changes in societal institutions and 'non-energy' policies, interact with patterns of consumption and energy demand over biographical time. Building on recent work, there is potential for bringing demand research into closer connection with accounts that draw biography into discussions of social reproduction and change. Scholars of reflexive modernisation have theorised about implications of institutional change in late modernity for everyday life. A key focus of this debate has been the consequences of the changing gender order and transformations in the female biography for domestic and relational practice. Røpke (2009) notes that these changes, in providing a backdrop against which transformations in performances occur, should be a focus of empirical analysis. However, to date they have remained under-explored in practice-theoretical research on consumption. In addressing this lacuna, the value of a biographic temporal scale for exploring "multiple dynamics already embedded in the social world" (Walker 2014: 49) emerges as particularly promising.

In response to these gaps, this chapter explores the implications of institutional transformations in the gender order for the demanding of energy in everyday life. Underpinning the discussion is one key question, that is, how do the everyday energy practices of individuals intersect and interact with processes of biographic, institutional and societal change? In seeking to provide an empirically driven account, this question is explored in the context of a recent biographic, qualitative investigation of energy demand conducted in Ireland. This study focused on understanding patterns and processes shaping biographic dynamics in key household practices implicated in energy demand, namely food, mobility and laundry practices. The term demand is used herein to denote consumption of energy as it is embedded in the performance of these socially recognisable and environmentally significant practices. In a European context, Irish exceptionalism in terms of the rate and pace of recent structural change offers a unique context in which to explore the processes under examination. The chapter begins by considering theoretical concepts sensitising the investigation. It then proceeds to provide an overview of the biographic-practice methodology and empirical study from which the data presented is drawn. Following this, findings are discussed in the context of the empirical light they shed on the research question. First, broad patterns in gendered practice careers are discussed. Following this, a more detailed exploration of the lives of two women is conducted to investigate in more detail the role of time and space in shaping the development of subjectivities and forms of practical action. Here attention focuses specifically on the implications of institutional and technological change for shifting temporal and relational dynamics of domestic practice. Knowledge about the changing allocation of practices in the day, as well as shifting relationalities of practice, in terms of who carries out practices and to what ends, is important for planning interventions to reduce energy demand. The chapter concludes by reflecting on the potential of biographic-practice approaches for exploring patterns of social reproduction in practice and advancing understandings of how socio-technical change intersects with dynamics in energy practices.

11.2   Considering Biography

Paths, Projects and Dialectics

In the context of this investigation, the work of geographer Allan Pred emerges as particularly significant. Pred's work (1981a, b) was specifically concerned with addressing a gap within practice-theoretical literature, namely how an individual's daily activity is reproduced or transformed over the course of their biography. To this end, he was concerned with exploring how the everyday activities, accumulated knowledges and biographies of individuals intersect and interact with the social reproduction and transformation of practices, societies and institutions.

Directing attention to the biography of individuals, Pred approaches the "time-space choreography of an individual's existence" through the concept of the path: "the biography of a person is ever on the move and can be conceptualised and diagrammed at daily or lengthier scales of observation as an unbroken, continuous path through time-space." (Pred 1981a: 9) Accordingly, an individual's life can be analysed at two scales of analysis: the 'daily path' and the 'life path'. The daily path refers to the activities, events and actions that take the individual through the time-space of their daily life, whereas the life path refers to the longer term and overlapping institutional roles in which they are associated over the course of the lives in domains such as family, work and other organisations.

In considering dynamics at these two scales, the path concept stresses the relational, material, spatial and temporal embeddedness of an individual's daily activity. In this respect, Pred highlights the "intricate interconnectedness of different biographies" as they merge and intersect in time and space (Pred 1981a: 10); as an individual makes their way through the world their daily path is constantly coupled and uncoupled with the paths of other individuals, technologies and natural objects, each with continuous biographies or time-space paths of their own.

At lengthier scales of observation, the composition of an individual's life path is shaped by their socio-historical socialisation and the number and mix of institutional roles in which their lives become enmeshed. Pred discusses the dialectical relationship between the longer term institutional roles and projects in which an individual is invested and the activities of their daily path, which he refers to as the "daily/life path dialectic". Institutional roles shape daily activity by configuring the types of projects and activities an individual is committed to; for example, an individual's participation in a work-related project moderates and constrains their ability to participate in projects relating to the self and family. Pred contends that it is at the intersection between an individual's path and the projects of institutions that the dialectic between agency and structure can be observed. How these intersections between institutional and personal temporalities shape patterns of demand in personal life is thus an important question for practice researchers.

Pred's concepts are useful for exploring dialectic processes in practice dynamics and hold potential for biographic investigations of energy demand. However, in emphasising the structuring of individual biographies, his account has paid less attention to individual agency and its transformation over time. In addition, while he discusses the role of socio-historical and institutional contexts in configuring individuals' paths, a more detailed consideration of the operation of political and normative forces in shaping lives is missing from Pred's account (Spurling 2010).

Institutions, Lives and Domestic Practice

Moving outside the practice-theoretical realm, other sociological accounts that bring biography into questions of social reproduction and change offer additional avenues for conceptualising the intersections between lives, practice and institutions. Accounts of reflexive modernisation (Beck and Beck-Gernshein 2002) have sought to theorise how dynamics in modern institutions connect and intersect with transformations in personal and relational life. Much of this debate has focused on changing gender orders and the implications for domestic life of transformations in female and male biographies in the context of institutionalised individualisation. According to these accounts, transformations in the female biography since the 1960s, associated with processes of institutional restructuring in the context of women's liberation and participation in paid labour, is resulting in the democratisation of relational life and a dismantling of traditional, institutionalised family practices and structures.

Gidden's and Beck's theoretical analyses highlight lifecourse arrangements as becoming less standardised and suggest an extensive undoing of traditional gendered and class-based ways of being. However, these accounts have been challenged empirically by Foucauldian and feminist-inspired analyses that highlight the continuation of structural inequalities in personal and working life (Wharton 2012). This work suggests that, far from being neutral, the reflexive biography is inherently bound up with difference, power and inequality. For example, despite the dramatic increases in female participation in the labour market in Ireland, recent time use data indicates that a highly unequal division of labour continues in Irish homes (cf. McGinnty et al. 2005). Thus, ambiguity exists in terms of the consequences of recent societal changes for the fabric of personal and domestic life, with some accounts emphasising radical discontinuities and others stressing the continuation of structural differences. Furthermore, these accounts have paid little attention to the materiality of social life. As such, the crucial role of technology as a mediator in social and biographic change has been overlooked.

In summary, accounts that bring biography into a discussion of social reproduction and change offer fruitful avenues for advancing understandings of dynamics of demand. While Pred's path and dialectic concepts sensitise the researcher to the embeddedness of individuals' daily activity in time, space and materiality, accounts of social change in late modernity direct attention to processes of continuity and change in relational lives in the context of the wider historical and institutional contexts in which the lives unfold. Bringing these accounts together in an analysis of the intersection between lives and practices offers a contextualised approach that considers materiality as well as social and institutional contexts. To date the relationship between wider socio-technical transformations and dynamics of demand has not been explored in an Irish context. The remainder of this chapter discusses an empirical investigation into the dynamics of demand in the context of the lives of Irish people.

11.3   Researching Biographic Dynamics in Energy Demand

The Research Context

In exploring the intersection between institutions, individual lives and everyday practice, the case of Ireland provides an interesting research context. Prior to its entry into the European Community (EC), Ireland was noted for its restrictive policies in relation to women's opportunities with discriminatory law regarding the legal status and employment of married women enshrined in the Irish constitution in 1922. The social policy known as the 'marriage bar' restricted the employment of married women in a host of key professions. Pressures on the government to meet growing gender equality standards climaxed in the context of Ireland joining the EC in the 1970s. The Irish government, keen to be seen to be meeting standards, commissioned the historic report on the Status of Women in 1970. Published in 1973, this report formed the blueprint for change, paving the way for the abolishment of the marriage bar in 1973 and the gradual advancement of women in Irish society (Owens 2005). How these social changes have intersected with the performance of environmentally significant domestic practices in Irish homes, however, has not been investigated from an experiential, biographical perspective. To this end, it is important to ask how these wider changes have intersected with relational and temporal patterns of energy demand.

A Biographic, Practice-Orientated Methodology

Within sustainability research, biographic methodologies (Chamberlayne et al. 2000: 2) have received growing attention as offering contextual, dynamic and experiential tools for researching demand (Jaeger-Erben 2013). In this investigation, an in-depth, qualitative, reconstructive-biographic approach was developed to shed light on contexts, experiences and dynamics of energy demand for food, laundry and mobility over individuals' lives. In cross-fertilising biographic and social practice approaches, the methodological design for this study combined various practice-oriented, narrative and lifecourse tools to represent and analyse the intersections between individuals' lives, their everyday practice and wider socio-historical contexts.

The methodology comprised three distinct yet interconnected stages (see Table 11.1). First, biographic-narrative interviewing involved eliciting detailed accounts of individuals' wider biographic history and careers of domestic practice. Lifecourse and energy practice (food, laundry and mobility) timelines complemented the interview, providing a

Table 11.1 Methodology

Data collection phase	Vertical analysis Life path-daily path dialectic Internal-external dialectic
Phase 1 Interviews *Life path*	*Zooming out* Constructing broader biography Constructing individual's life pathways and practice careers Timelines, practice graphs
Phase 2 Diary (2 weeks) *Daily path*	*Zooming in* Constructing daily path practice dynamics Material, temporal, social and spatial contexts of everyday practice
Phase 3 Interviews House tours *Life path, daily path*	*Zooming in and out* Constructing daily path practice dynamics at different life stages Constructing individuals' life path careers

visual-descriptive reconstruction of individuals' lives and facilitating a detailed discussion of the intersections between biographic context and the practices under investigation. Following this, participants took part in a two-week semi-structured practice diary task in which they recorded information on their energy practices. Structured sections requested standardised information from participants on the temporal, relational and material contexts of their food, laundry and mobility practice. Finally, follow-up interviews involved further exploration of participants' routines at previous life stages as well as transformations of their practice over time.

11.4　Sample

Biographic research commonly employs a theoretical sampling format, whereby the aim is not to produce representative results but to gain an in-depth, experiential and contextual understanding into a phenomenon with a small number of participants (Jaeger-Erben 2013). A theoretical sampling format sought the recruitment of 14 participants with diverse views and life-course circumstances. The sample criteria included age/

Table 11.2 Sample

Born	1925–40	1941–55	1955–70	1970–85
Male	Tony (75)	Henry (69)	Michael (59)	Daniel (47)
		Frank (67)	Seamus (58)	
			James (56)	
Female	Billie (87)	Alison (62)	Bridget (56)	Sara (46)
	Grace (81)	Martha (61)	Triona (51)	

birth cohort, gender, family structure, class, location and reported levels of environmental concern and practice. The rationale for this emphasis on diversity was to explore the role of time and space in configuring demand as well as to expand the range of insight into practice dynamics in general. Field work took place over a one-year period. See Table 11.2 below for an overview of participants by gender and age cohort.

11.5   Exploring the Intersections Between Lives, Institutions and Practice

Gender and Age Structured Practice Careers

Analysis revealed that long-term trajectories of career development in the domestic practices were closely linked to the institutional roles and pathways individuals pursue, with these being heavily structured according to gender and cohort. The gender and age differentiation of lifecourses was especially evident in terms of trajectories and roles within the domains of family and employment. In terms of the dominant fields structuring individuals' daily lives, men's lives and social identities were strongly orientated around institutional work roles. In contrast, women's working careers were truncated with long periods of absence and their roles were overwhelmingly orientated around the domestic sphere and family relationships. Women's increasing participation in institutional roles and projects outside the domestic sphere was the most pronounced transformation to emerge, with younger women in the sample more likely to continue working throughout their married life. In contrast, none of the women in the oldest cohort worked following marriage. Significant structural and institutional changes following Ireland's entry to the EC, most notably a questioning of structural patriarchy, the removal of the marriage bar and technological development, were noted by participants as key factors re-configuring opportunities for Irish women over time.

The intersection of life paths with institutional change shaped dynamics at the scale of the daily path in terms of individuals' work career trajectories as well as the temporalities and relationalities of domestic energy practice. With regards to work career development, the gendering of lifecourse roles was reflected in the energy-consuming practice careers of men and women. Men were far less likely to have developed continuous careers in domestic practices associated with the private domain. Younger men were more likely to have developed careers in domestic practices such as food preparation, laundry and cleaning whereas men in older generations developed these careers later in their lives or not at all. Periods of non-participation, that is phases in a career denoting absent or very infrequent participation, were evident in all men's food and laundry careers (see Table 11.3). Men's participation in these practices was overwhelmingly shaped by events in the life of their spouse, with patterns of non-participation tending to be reproduced until they were disrupted by an event in the female spouse's life, such as returning to work or illness. Participants in younger cohorts were more likely to have a female spouse working. However, despite this their wives still assumed the bulk of responsibility for these private domain domestic activities.

Conversely, all women in the sample reported as proficient and regular practitioners in domestic practices such as food preparation, laundry and

Table 11.3 Men's participation in food and laundry over the life path

Life stage	70s Tony	60s Henry	60s Frank	50s Michael	50s James	50s Seamus	40s Daniel
Childhood	NP	NP	NP	P	P	P	NP
Young adulthood	NP	NP	NP	P	P	P	P
Married life	NP	NP	NP	P	NP	NP	NP
Later married life	NP	SP	NP	P	SP	SP	SP

Legend: *NP* non-participation, *P* full participation, *SP* small participation

Table 11.4 Instances of performance per week by gender

| | Mean | |
Food	Female	Male
No. grocery shops	4	1.3
Time on food prep in mins	296	118
Meals prepped for self	15	8
Meals prepped for others	10	1
Meals prepped by others	2	3
Pre-prepped meals	2	1
Laundry		
Performed by oneself	2.5	0.85
Performed by another	0.3	3

cleaning, with this pattern developing over a life time of socialisation into gendered roles. Across all age groups, diary data revealed that women continue to hold the bulk of responsibility for domestic practices, and carry a far greater number of practices for other people (see Table 11.4). In contrast, men were likely to carry out practices for themselves, with fewer tendencies to perform practices for others on top of that. In addition, they were much more likely to report having domestic practices, including shopping, food preparation and other domestic tasks such as cleaning and laundry, performed for them, in each case by their female spouse.

However, despite the continued feminisation of domestic labour, younger women were more likely to report a degree of participation by their spouses. In this regard, work commitments played a crucial role. Houses that had a female spouse working indicated differential relational and temporal organisation of practices in their homes. In terms of the temporalities of practice, domestic practices were plotted in 'hot spots' around the institutional time structure of their work routines. The role of technology in enabling these changes emerged in participants accounts. Narratives revealed that the technologisation of practice has shifted the temporalities of practice, with practices taking less time within the day but being of increasing resource intensity. These shifting temporalities were a crucial element facilitating the realisation of non-energy policies and priorities in relation to women's increasing participation in work roles (Table 11.5).

Table 11.5 Instances of performance per week by gender and age

	Female mean		Male mean	
Food	<65	>65	<65	>65
No. grocery shops	5.5	3.5	1.6	1
Time on food prep in mins	326	265	112	123
Meals prepped for self	17	12.5	5	9
Meals prepped for others	8.5	11.5	0.5	1.5
Meals prepped by others	2	3	7	6
Pre-prepped meals	0	4	0	1
Laundry				
Performed by oneself	1.5	3.5	1.6	1
Performed by another	0	0	3	3

A differentiated picture emerged in relation to car driving careers; women in younger generations were more likely to be recruited to car driving earlier, usually in their 20s, whereas women in older generations were recruited later, in their 30s and 40s, or not at all. In contrast, all men in the sample were recruited to car driving in their teens and 20s. These patterns reflect broader patterns in the expansion of driving in Ireland, reflected in the growing number of cars and licenses per capita (see Table 11.6).

It would appear that the recruitment of women to car driving has brought about an overall increase in energy consumption. Prior to their recruitment, many older women combined occasional car trips as passengers with lower resource-intensity mobility modes, with practices of walking and using the bus more prevalent, often performed daily, and embedded within the rhythms of daily living. Furthermore, children were more likely to make the commute to school using low energy transport modes, such as walking, cycling or using the bus. However, following women's recruitment, patterns of multi-modality among both women and children declined, with practices such as the commute to school and food shopping reported as increasingly performed with the car.

The patterning of individuals' practice according to generation and gender brings into relief the role of socio-historical context in shaping or foreclosing the development of different types of subjectivities and forms of practical action. The changing configuration of the daily-life path dialectic over cohort groups indicates a broader pattern of transformation in

Table 11.6 Cars and licences per capita

	1960	1970	1980	1990	2000	2010	2014
Irish population	2,832,100	2,949,900	3,401,000	3,505,800	3,789,500	4,554,800	4,609,600
Total cars	169,681	384,273	646,609	796,408	1,314,059	1,872,715	1,943,868
Cars per capita	0.06	0.13	0.19	0.23	0.35	0.41	0.42
New cars licenced	27,941	52,947	91,728	83,420	225,269	84,907	92,361
New cars per capita	0.01	0.02	0.03	0.02	0.06	0.02	0.02

	1993	2000	2008				
Passing driving test	52,062	93,315	222,145				
Population	3,574,100	3,789,500	4,485,100				
New licences per capita	0.01	0.02	0.05				

biography shaped intersecting processes of institutional and technological change. In exploring these processes in greater depth, the domestic practice histories of two women from two different generational cohorts are presented and discussed below.

11.6   Reproduction and Change in Routines and Practice

Billie's Daily Path

Billie is an 87-year-old mother to 4 adult children and 12 grandchildren. She is married to husband Jack, also 87 years. The couple continues to live independently in their suburban home in Galway. Marked gender divisions in practice exist between Billie and her husband; Billie looks after all of the indoor domestic practice such as cleaning, food preparation and cooking, while Jack never participates in these tasks, instead looking after more 'masculine' jobs such as DIY jobs. Until recently Jack drove the couple to the supermarket twice a week to collect groceries. However, he has stopped driving due to declining sight. Billie never learned to drive and so now relies on family members to bring her shopping once or twice a week.

Martha's Daily Path

Martha is a 61-year-old mother to five sons and grandmother to six grandchildren. She lives with husband James in a suburban home in Dublin. Martha works 30 hours a week in an administrative role and she and James share many of the household domestic duties. James is a proficient cook and prepares most of their meals, while Martha looks after most of the laundry and cleaning activities. Both Martha and James own and drive their own cars, which Martha contends they would be unable to do without.

At face value, both women share comparable lifecourse circumstances: They are both grandparents, live alone with their spouses, and enjoy a stable home and family life. However, the allocation of practice in Billie's home is marked by segregated gender roles, whereas in Martha's case a more equal sharing of practices is observed. In explaining these differences, a consideration of how their current daily practices are rooted in biography is needed. Of interest to this investigation are the intersections between structure, life opportunities and the configuration of everyday practice in the home. In this respect, qualitative differences became evident in the women's biographical narratives.

Biographic Pathways and Domestic Practice

Comparative analysis of the women's lives revealed similarities and differences in life pathways, personal concerns and projects pursed. Both women married and became mothers in their 20s. However, socio-historical timing emerged as highly significant in configuring differential opportunities for the development of a career outside the private domain. While institutional and normative changes in relation to women's liberation emerged to shape Martha's pursuit of a work career later in her life, for Billie these opportunities were not emphasised within her narrative or pursued in her life trajectory. As discussed below, these differences in opportunity emerged to shape the configuration of domestic practices in the women's homes.

Throughout their childhoods, Billie and Martha experienced socialisation into gendered domestic practice roles. Within both of their homes, practice was segregated according to gender divisions and their accounts of socialisation emphasised their early enrolment into domestic practices by their mothers. During their 20s both women got married and, in accordance with the customs of the time, forfeit their jobs and became full-time homemakers. However, while Billie was more than content to fall into this role, stressing that she *"didn't mind at all"* and *"enjoyed every bit of it"*, Martha emphasised the restrictions placed on women in the context of the marriage bar:

> When I got married in the early 70s, girls gave up work – they still had to give up. It wasn't really acceptable to still work as a married wife or mother

then… things changed after the marriage bar was removed, not straight away though, it took time for people to accept women having a life outside the home…

Following marriage, the women's daily paths were reconfigured around their new roles as wife and full-time homemaker and marked gender division in roles and practices were reproduced, with the wives' positions firmly located in the private sphere and their husbands operating in the public sphere as male breadwinners. While for Billie this pattern was reproduced throughout her whole lifecourse, for Martha a significant transitional period emerged later in her life in which this pattern of reproduction was disrupted and challenged. In the context of a changing normative landscape, the actualisation of an independent identity outside of the family was now a socially accepted possibility and was increasingly foregrounded by Martha as an important personal goal. In the early 1990s she decided to return to work:

…as the boys grew up I started to feel that I needed to find myself as an individual again…. I made the decision to go back to work. I found a great sense of freedom in that….to find my own identity again without being identified as being someone's wife, someone's mother…

With regard to the configuration of practices with her home, the event of Martha returning to work stimulated a process of transformation in terms of how, when and by whom domestic practices were carried out (See Fig. 11.1). This process had path dependent implications that emerged and developed over the following decades.

Martha's commitment to the new work role resulted in a realignment of her daily path around the institutionalised timetable. The temporality of Martha's domestic practice changed. She began pre-prepping and freezing food rather than cooking every evening. In addition, her domestic practices become funnelled into 'hot spots' around her work timetable. In contrast, the temporalities of Billie's practice remained much more stable throughout her career. Practices of pre-prepping did not feature in her account in a regular way and the temporal ordering and sequencing of her domestic practices were extended throughout the day.

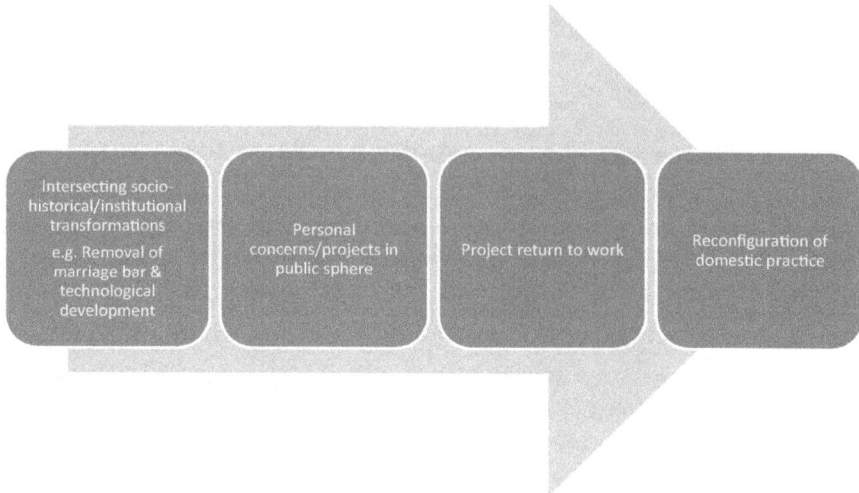

Fig. 11.1 Sequence of causality shaping configuration of Martha's practice

With regard to the relationalities of practice, these changes in Martha's daily path created a context in which James was, for the first time, required to participate in domestic activities. He gradually took on the task of preparing pre-cooked meals, food shopping and performing some laundry activities. James' process of learning via participation and trial and error occurred incrementally, taking on a path dependency that operated over many years. Eventually, with the help of technologies such as the slow cooker and microwave, he began cooking meals himself. Martha discussed the incremental changes as follows:

> When I started working, I was still doing it all... housework would be done, they'd (children) all have been fed, and the bacon would be cooked and all he would have to do is dump the cabbage into it and everything else would be ready to go... But by the time that year was out James was coming in and the potatoes mightn't have been ready and he might have to cook the pork chops or something like that for them. That's when it started and gradually he did more... Before that he did absolutely nothing. Growing up his sisters were told, James would like a cup of tea, that kind of thing... No, he never cooked or house worked or anything like that... I would have done 99 percent of it before... But by the time that year was

out he was coming in and cooking... we got one of these slow cookers... and he actually got quite interested in recipes and stuff for that and used the internet to... get recipes... so it's been a huge change when you think of it really

Martha's narrative reveals the interconnection of personal and material biographies in the reconfiguration of the temporalities and relationalities of daily practice. Technologies played a crucial mediatory role facilitating the shifting temporality and sequencing of her domestic practice and enabling the gradual transfer of practices from the more skilled practitioner, Martha, to the less skilled practitioner, James. While James was unskilled in food preparation, the slow cooker enabled him to participate in the practice of cooking, accounting to some degree for his lack of biographically accumulated skill. The role of technology in freeing up women's daily paths was evident in all cases analysed but was most apparent in cases where a female spouse pursued a work career outside the home. The increasing technologisation of daily practice has direct implications for the rising resource intensity of practices. Technological appliances were crucial in enabling couples to manage the temporal constraints placed on their daily paths by the joint institutional commitments and in some instances for facilitating men's participation. In Martha's case, the result of this process was a gradual destabilisation and reconfiguration of the traditional gender role setting in the home, and, over time, a more equal sharing of practices between Martha and James transpired.

This necessarily abbreviated overview of Billie and Martha's cases highlights the importance of considering biography in understanding the configuration of current day routines. By connecting the women to their respective cohorts and grounding their accounts in rich and nuanced empirical evidence, the role of time and space in creating or foreclosing circumstances for the development of practice careers (lifecourse and domestic) is revealed. The technologisation of domestic energy practices, such as food and laundry, have gradually steered practices towards increasing resource intensity. This process has been a crucial perquisite for enabling women to be released from the burden of temporally and physically demanding housework and for institutional change regarding gender policies to transpire. In this respect, it is clear that energy demand is in part the unforeseen consequences of 'non-energy' policies and priorities (cf. Butler et al. 2016; Royston 2016), such as women's participation in work and changing gender roles and norms. In shaping 'who does or does not have access to specific domains at specific times to do specific things' (Pred 1977: 208), institutions shape the actions and event sequences of an individual's daily existence. Techno-material transformations associated with the 'creeping technologisation' of everyday practice (Henwood et al. 2015) provided the context in which institutional transformations in gender policies could be realised in everyday performance. In this respect, intersecting technological and institutional change played a crucial role in shifting the tempo, rhythm and relational patterning of energy demand at the daily path. The intersections of institutions, practices and lives can be explored by focusing on the nature of the relations that link them and how these dynamically evolving relations work to enable and constrain ways of acting over time.

11.7   Conclusion

This chapter set out to explore how the lives of individuals, practice and institutions intersect. The data reveals that patterns of career development in energy practices are closely linked to the institutional roles and pathways individuals pursue, with these being structured according to gender and socio-historical location. Furthermore, the interconnectedness of material and institutional transformations in creating or foreclosing new temporalities and relationalities of performance of environmentally significant domestic energy practices, such as those relating to food and mobility, was revealed. In considering the implications for energy demand of the changing practice careers of the men and women, technological change played a crucial role in shifting the tempo and rhythm of energy demand at the daily path, facilitating women's participation in new work roles and men's recruitment to practices previously performed almost exclusively by women. Further, changes associated with the technologisation of daily life have resulted in increasingly resource-intensive practice arrangements, upon which contemporary domestic and work schedules depend. Understanding the shifting temporalities and relationalities of practice that have emerged from recent socio-technical change is important in informing the design of interventions to reduce energy demand and consumption.

The implications for domestic life of women's participation in paid work has received academic attention for several decades (Hochschild 1997; West and Zimmerman 1987). However, these accounts have paid little attention to the materiality of social life and in doing so have overlooked the role of technology as a mediator of change. Furthermore, to date empirical investigations in an Irish context have been lacking. By bringing together practice-theoretical and sociological accounts that consider biography into discussions of social reproduction and change to explore dynamics of demand in a new empirical context, this chapter has sought to highlight the value of a biographic-practice approach for exploring interacting processes and mechanisms shaping dynamics in energy demand. In facilitating an examination of incremental processes and sequences of causality that often remain eclipsed as they occur in situ, biographic methods offer fruitful avenues for investigating the interplay of structure and agency operating over extended temporalities. Combining an analysis of patterns across generations with experiential accounts of individuals' lives, the findings shed light on the institutional and infrastructural ordering of lives and practices.

An important contribution of a biographic-practice approach is that it facilitates investigation into broader contexts and processes which frame and unintentionally shape dynamics in performance. Retrospective approaches that proffer extended temporal frames for analysis offer a means of exploring the role that governing processes and policies play in steering social action. However, these methods are not without their limitations. Caution must be taken when making generalisations based on small, purposive samples especially in relation to what analysis can say about the role of context. For example, instances of women working outside the home even before discrimination was illegal have not been studied here. However, despite these findings not being generalisable in a statistically representative sense, following the practice-biographic theoretical framework, it is maintained that theoretical insights into the operation of social processes can be revealed by exploring the realm of the experiential and performative. As such, in seeking to uncover contextual processes, the findings aim for theoretical generalisability, stressing the value and relevance of theoretical insight, over the representation of a population.

In summary, in considering the key challenges of representing multiple, intersecting scales and temporalities in practice dynamics, this chapter has suggested that practice-theoretical and social science concepts and methods that have engaged with the question of biography hold potential. These fields offer a largely untapped body of conceptual and methodological tools that could be usefully deployed in future demand research, exploring energy, mobility and food, to capture, represent and understand biographic-practice dynamics in ways that complement other recent theoretical and methodological innovations in this rapidly evolving field.

Acknowledgements This research received funding from the Irish Research Council Postgraduate Scholarship Scheme (2013–2016). The author also wishes to express thanks to all participants who were involved in this research, which was conducted with a sample who participated in the CONSENSUS Project: 'A cross border household analysis of consumption, environment and sustainability' (http://www.consensus.ie), funded by the Environmental Protection Agency (EPA) Science, Technology, Research and Innovation for the Environment (STRIVE) Programme 2007–2013.

Bibliography

Beck, U., and E.B. Beck-Gernshein. 2002. *Individualization: Institutionalized individualism and its social*. London: Sage.

Burningham, K., S. Venn, I. Christie, et al. 2014. New motherhood: A moment of change in everyday shopping practices? *Young Consumers* 15 (3): 211–226.

Butler, C., K. Parkhill, and K. Bickerstaff. 2016. *Welfare policy, practice and energy demand*. Demand international conference, Lancaster, 13–15 April.

Chamberlayne, P., J. Bornat, and T. Wengraf. 2000. *Turn to biographical methods in social science: Comparative issues and examples*. London: Routledge.

Greene, M., and H. Rau. 2016. Moving across the life course: The potential of a biographic approach to researching dynamics of everyday mobility practices. *Journal of Consumer Culture* (in press).

Hards, S. 2011. Careers of action on climate change: The evolution of practices throughout the life-course. In *Social policy and social work*. York: University of York.

Henwood, K., N. Pidgeon, C. Groves, et al. 2015. *Energy biographies research report*. Cardiff: Cardiff University. Available at http://energybiographies.org

Hochschild, A.R. 1997. *The time bind: When work becomes home and home becomes work*. New York: Metropolitan Books.

Jaeger-Erben, M. 2013. Everyday life in transition: Biographical research and sustainability. In *Methods of sustainability research in the social sciences*, ed. F. Fahy and H. Rau. London: Sage.

Jaeger-Erben, M., and U. Offenberger. 2014. A practice theory approach to sustainable consumption. *Gaia*, 23: 166–174.

Lanzendorf, M. 2010. Key events and their effect on mobility biographies: The case of childbirth. *International Journal of Sustainable Transportation* 4 (5): 272–292.

McGinnty, F., H. Russell, J. Williams, et al. 2005. *Time-use in Ireland 2005: Survey report*. Dublin: The Economic and Social Research Institute.

Owens, R.C. 2005. *A social history of women in Ireland, 1870–1970: An exploration of the changing role and status of women in Irish society*. Dublin: Gill & Macmillan.

Pred, A. 1977. The choreography of existence: Comments on Hägerstrand's time-geography and its usefulness. *Economic Geography*, 53 (2): 207–221.

———. 1981a. Social reproduction and the time-geography of everyday life. *Geografiska Annaler. Series B, Human Geography* 63 (1): 5–22.

———. 1981b. Of paths and projects: Individual behavior and its societal context. *Behavioral Problems in Geography Revisited*: 231–255.

Røpke, I. 2009. Theories of practice—New inspiration for ecological economic studies on consumption. *Ecological Economics* 68 (10): 2490–2497.

Royston, S. 2016. *Invisible energy policy in higher education*. Demand centre conference, Lancaster, 13–15 April 2016.

Sayer, A. 2013. Power, sustainability and wellbeing: An outsiders view. In *Sustainable practices: Social theory and climate change*, ed. E. Shove and N. Spurling, 167–180. London: Routledge.

Shove, E., and M. Pantzar. 2007. Recruitment and reproduction: The careers and carriers of digital photography and floorball. *Human Affairs* 17 (2): 154–167.

Shove, E., M. Pantzar, and M. Watson. 2012. *The dynamics of social practice: Everyday life and how it changes*. London: Sage.

Southerton, D. 2006. Analysing the temporal organization of daily life: Social constraints, practices and their allocation. *Sociology* 40 (3): 435–454.

Spurling, N. 2010. *Authors of our own lives? Individuals, institutions and the everyday practice of sociology.* Lancaster: Department of Sociology, Lancaster University.

Walker, G. 2014. The dynamics of energy demand: Change, rhythm and synchronicity. *Energy Research & Social Science* 1 (0): 49–55.

Warde, A. 2005. Consumption and theories of practice. *Journal of Consumer Culture* 5 (2): 131–153.

West, C., and D.H. Zimmerman. 1987. Doing gender. *Gender & Society* 1 (2): 125–151.

Wharton, A. 2012. *The sociology of gender.* Oxford: Wiley-Blackwell.

Mary Greene is a PhD researcher at the School of Geography and Archaeology, NUI Galway, Ireland. Her research interests focus on the dynamics of everyday practices over time and space and are situated at the interface between human geography, environmental sociology, environmental psychology, lifecourse studies and science and technology studies. She is especially interested in the interrelationships between socio, cultural and technological change and evolution of social practices in peoples' everyday lives. Her doctoral research situates these dynamics biographically to explore how practices change over the lifecourse within changing socio-technical landscapes.

12

Demanding Business Travel: The Evolution of the Timespaces of Business Practice

Ian Jones, James Faulconbridge, Greg Marsden, and Jillian Anable

12.1 Introduction

Compare walking down the hallway to collect a report from a printer on the way to a client meeting in an office in the same city with collecting the report on the way home as part of a larger set of activities to prepare for an international flight to meet a client in a different country. The energy demand of the former is a number of magnitudes less than the latter. Further, in addition to environmental implications (Lassen 2010; Urry 2012) differences in travel that arise from the two examples also have important implications for the well-being of workers (Gustafson 2006, 2012) and corporate finances (World Travel and Tourism Council 2011). Yet, despite these effects, there continues to be a growing tendency for the doing of business to involve travel, very often international. How do we explain this?

I. Jones (✉) • G. Marsden • J. Anable
Institute for Transport Studies, University of Leeds, Leeds, UK

J. Faulconbridge
Lancaster University, Lancaster, UK

© The Author(s) 2018
A. Hui et al. (eds.), *Demanding Energy*, DOI 10.1007/978-3-319-61991-0_12

As part of the mobilities turn (Sheller and Urry 2006), significant effort has been made to analyse the relationships between the globalisation of professional services (such as accounting, advertising, architecture, engineering and law) and 'portfolios' of business mobility which produce an internationally hyper-mobile class of workers (Faulconbridge et al. 2009; Millar and Salt 2008; Salt and Wood 2012). To date, the majority of the business travel literature has focused on charting types of travel and the face-to-face contact it allows (e.g. Davidson and Cope 2003; Faulconbridge et al. 2009; Jones 2007; Lyons 2013; Millar and Salt 2008). This, we contend, neglects the important question of how demand for business travel, and what it enables (e.g.: face to face, sales, staff management), comes to be ingrained in how business is done.

In this chapter we offer a new perspective on demand for travel by focusing on the rise of global professional firms; specifically global construction and engineering consulting firms. We do this by drawing upon Schatzki's (2006b, 2009) notion of timespace. Specifically, we explore how demand for business travel arises as a result of the need for the coordination of business through interwoven timespaces. This view is found to offer new insights into the creation of and ways of thinking about the challenge of reducing demand for business travel. In particular, it reveals how forms of corporate organisation have co-evolved with the role of mobility in society. We argue that over time, changes to the interweaving of timespaces have evolved in ways inseparable from the possibilities for and provision of business travel; and in turn this creates a contemporary situation of significant and hard to reduce demand for travel.

We begin by summarising insights and knowledge gaps from an emerging body of work that has traced the reasons for business travel. This is followed with a summary of our case study of two UK consulting firms. Schatzki's notion of timespace is then introduced and applied to interpret a contextualised history of how coordination is achieved in professional service firms when procuring and arranging work. We conclude with a wider discussion of the value of our approach for understanding mobility-intensive business practices.

Conceptualising Demand for Business Travel

Research has identified how business travel facilitates attending meetings (Lyons 2013) and conferences (Storme et al. 2016); working on global projects (Faulconbridge 2006); managing corporate subsidiaries and those working in them (Jones 2007); attending to existing clients (Millar and Salt 2008) and identifying new ones (Wickham and Vecchi 2009). Underlying such analyses is recognition of the 'compulsions of proximity' (Urry 2003), and in particular the continued significance of embodied encounters (Strengers 2015) in the doing of business. Such concerns speak to the teleo-affective nature of travel. Teleology highlights goals and ends of a particular practice (e.g. attending to existing clients), whilst affectivity highlights a range of emotions that inform the doing, and/or reflect what is acceptable (e.g. social expectation of co-presence when pitching an idea to a new client) (Schatzki 2005, 2006b, 2012b). As a result, the 'need' for travel is most often explained by what it enables, what cannot be achieved at a distance, and what is expected in the corporate world.

Reducing demand for business travel is, however, a widespread ambition. This stems from environmental concerns—how companies might reduce travel and its carbon impacts—as well as a desire to reduce the cost of travel and negative effects upon employee well-being. Such concerns are part of a long running debate about substitution and the role of, in particular, video conference technologies. To summarise, questions are raised about what the 'bandwidth' of tele- and screen-based encounters cannot facilitate and how technological developments might recreate the affordances of 'being there' (e.g. Arnfalk and Kogg 2003; Gallié and Guichard 2005). The main response to such debates has been to emphasise the limitations of substitution, given the way that physically inhabiting the same space brings about opportunism (Storme et al. 2016) and corporeal encounters (Strengers 2015) that are valued and many do not want to forgo. Hence co-operation (Faulconbridge et al. 2009) and alliance (Haynes 2010) between travel and technologically facilitated interactions has been highlighted. Workers and firms, therefore, find themselves in the difficult situation of wanting to reduce travel for a variety of reasons, but being unable to imagine conducting business without mobility (Cohen 2010; Lassen 2010).

In this chapter we treat the demand for business travel as arising from the ongoing reproduction of bundles of spatially stretched work practices (Shove et al. 2014; Thrift 2008). These practices involve a range of tasks, activities and decisions (henceforth collectively referred to as actions) that are coordinated through the interweaving of timespaces. Demand for business travel, therefore, is suggested here as not simply the product of function, like the need to physically travel to touch and fix a photocopy machine (Orr 1996). Nor is it the product of conventions around building trust through co-presence (Jones 2007). These factors undoubtedly encourage travel. However, we argue that the root cause of demand for travel is the development of forms of corporate organisation that necessitate the interweaving of the timespaces of spatially distributed and temporally dispersed actors so as to coordinate a range of actions. We argue that over time interweaving has become more complex as forms of organisation have co-evolved with changing information and communication technology (ICT) and mobility affordances and provision. This implies that demand reduction might arise not from virtual ways of working, but through corporate reorganisation that produces work practices which require the interweaving of the timespaces of less spatially stretched actor constellations, or interweaving not premised on travel.

The Case and Approach

Our starting point for capturing changes in demand for business travel is to place business practices centre stage (Reckwitz 2002; Shove et al. 2012). In part we are inspired by Jones (2013: 63) who suggests that business travel itself might be conceived of as a "socio-economic practice that involves individuals or groups moving for the purposes of economic activity". For Jones (2013), the benefit of this focus comes from the three interrelated insights gained from understanding what people do: spatio-temporality is revealed, in terms of business travel's effects on time and space; function becomes clear in terms of economic outcomes; and analytical clarity is gained when travel is significant and contributes to economic outcomes. This focus suggests understanding how travel correlates with company strategy and spatial work arrangement (Aguilera 2008) and helps to achieve business outcomes (Jones 2013).

We agree with Jones's call to adopt a practice perspective. However, there is one key consideration on which we diverge. In viewing business travel itself as a practice we learn a lot about the affordances of travel, but little about the underlying sources of demand for travel. Suggesting that the need to meet face-to-face is the cause of the demand for travel relies too heavily on a teleo-affective account of co-presence, providing limited insight into why corporate organisation has evolved in a way that makes it reliant on travel-enabled teleo-affectivities. In the engineering and construction business sector, like many business sectors, overcoming spatial constraints has, and continues to be, a central defining feature of recent forms of corporate organisation. This, we contend, is the source of demand for business travel, and hence if the intention is to better understand sources and in turn ways of reducing demand, the work practices associated with contemporary forms of corporate organisation and the way their coordination is achieved through interwoven timespaces should be the focus of a practice perspective. To develop this argument, we draw from a one-year case study of how two professional service firms procure and arrange work.

The two UK-based firms work in the civil engineering and survey consulting business sector. This sector provides design and construction services to a range of public and private clients. The fundamental work of these firms is the application of specialist technical knowledge such as engineering, architecture and planning to the design and construction of a range of waste, transport and energy projects. Essentially, firms procure work from a client (e.g. private/public rail sector), for a project (e.g. construction of new or alteration of existing piece of railway) which then needs to be delivered by a team within the firm. Both of our case study firms have a strong UK presence, but also have a global reach and reputation. The firms were selected by obtaining a list of candidate firms from the Guardian UK 300. Based on rankings of the top ten civil engineering and consulting firms over the past five years, a list of potential firms was identified and access gained to two through existing professional connections.

The limitations and strengths of conducting a case study of a phenomenon are well documented (Bryman 2008; Yanow and Schwartz-Shea 2006). Though lacking in generalisability, case studies can reveal a richer, more nuanced depiction of a phenomenon. The chosen case study is valuable for a number of reasons. First, the firms' global reach allows us to develop new knowledge about the creation, maintenance and suppression of international business travel, which is recognised as a particular environmentally damaging and personally demanding form of mobility. Second, compared to finance, IT and mining, the construction and engineering consulting business sector has received significantly less attention by scholars, and thus we offer new insight into demand in a travel-intensive sector (Anable et al. 2015). Third, our two firms have successfully maintained their global presence through continued firm consolidation and reorganisation, most recently in light of the Global Financial Crisis. This provides a window on how the evolution of corporate organisation affects demand for travel.

Twenty one-to-one interviews, based on a semi-structured interview protocol, were conducted, transcribed and then coded and analysed through QSR NVivo software. After an initial introduction into each case study firm by professional contacts, participants were identified by snowballing (i.e. feedback from interview participants) and criteria-specific identification (i.e. from review of corporate documents) (Bryman 2008; Neuman 2011). Interviews generally lasted one hour, beginning with participants describing their day-to-day tasks, and then shifting to discussions of procuring and arranging work. Rather than ask why virtually mediated ways of working don't substitute for physical ways of working or why workers travel, we sought to draw out how demand for travel arises from coordinating actions involved in procuring and arranging work in a globalised business world.

Business Practices and Travel Demand

The globalisation of our two case study firms, like in other studies of business travel, is central to the story of the 'need' for travel. The speeding up of travel has extended the spatial reach of corporations and enabled their globalisation (Dicken 2011). This results in what Harvey (1989) refers to as 'time-space compression', where markets stretched across space are serviced largely through mobility. Time-space compression thus enables and encourages the creation of business practices which create and stabilise demand for business travel. Schatzki (2009, 2012a, b), through the concept of timespace, offers a valuable way of understanding the temporal and spatial features of the business practices that make up our case study firms and which are fundamental to demand for travel.

For Schatzki (2012b), timespace is a fundamental influence on action. It is composed of both objective and relational dimensions.

Objective time[1] relates to measurable units, such as the minutes and hours of clock time, whilst objective space relates to identifiable locations, in our case for example an office (in a city with a GPS location). Relational time, which can be referred to as temporality, is the result of past, present and future orientations which influence action (see Hydle 2015). Relational space, or spatiality, "embraces arrays of places and paths anchored in entities, where a place is a place to perform some action and a path is a way among places" (Schatzki 2012a: 19). It is the amalgamation of the locations and paths between them that together constitute the spatial realm of the actors in question. In timespace, objective and relational space and time are viewed as "inherently related constituted dimensions of action" (Schatzki 2012b: xi).

Schatzki draws attention to how coordinated action is only possible when timespaces are interwoven (Schatzki 2012b: 87). Interweaving relates to the way "lives hang together" (Schatzki 2012b: 66) as a result of forms of interdependence between different actors and their practices. Timespace can be interwoven when action occurs in the: same place, same time; different place, same time; different place, different time; or same place, different time. The key is that the actors are interdependent, and in turn coordination results. This means that actions "combine to achieve a result that someone intends to be achieved" (Schatzki 2012b: 69). Schatzki (2012b: 66–67) argues that interweaving and thus coordination occur in four ways: interpersonal structuring (shared ends, means, emotions), chains of action (each actor following/reaction to another), intentional directedness (when an actor is focused on the actions and emotions of another), and the medium of settings (connections between actors that give them shared lives, events and stimuli). As we discus more fully later, the need for interwoven timespaces generated by contemporary forms of corporate organisation means chains of action and intentional directedness are key in the firms in our case study, thus creating demand for travel.

Specifically, whilst coordination in our case study firms, like in all forms of economic organisation, "rests on common, shared and orchestrated timespaces", particularly important are common and shared timespaces because "orchestration, by itself, is unable to ensure coordination" (Schatzki 2012b: 71). Orchestration is limiting because it involves actors pursuing their own aims, despite these aims ultimately being interdependent with the aims of others (Schatzki 2009: 42). For our case study firms this risks undermining global corporate strategies and the desire to provide global 'one stop shop' services to clients. Common timespaces are more productive because they are characterised by "enjoined" actors who share "ends, purposes, motivations, places, paths" (Schatzki 2012a: 20). That is, actors work in an enjoined way to deliver global corporate strategy and service. Shared timespaces are also characterised by "the same ends or motivations or at the same places and paths", but actors are not enjoined and as such co-exist rather than collaborate (Schatzki 2012a: 20). Demand for business travel arises from the need to create interwoven, shared and ideally common timespaces that encompass spatially stretched actors procuring, producing and delivering services. Travel is also related to the desire for harmonised timespaces in which "the actions of people who are simultaneously proceeding in the same or connected settings smoothly fit together through the contingent adjustment of each person's behaviour to what others are doing" (Schatzki 2009: 43–44). Harmonised timespaces avoid conflict (conflicting timespaces occurring when different actors jostle rather than align their actions) and, as we discuss, are the panacea that our case study firms strive to develop.

The interweaving of common, shared and harmonised timespaces, which enable firms to procure and arrange work, "prescribe and circumscribe spaces, times, sequences and formal relations" that give character, rhythm and form to everyday life in organisations (Schatzki 2012b: 141).

As far as business travel is concerned, this means considering how the need for the coordination of actions accomplished through different interwoven timespaces creates demand for business travel. With this goal in mind, we begin by tracing important changes in the sector over the past century. Drawing from secondary sources (pre-2000), the contextualised historical account provides the backdrop for understanding the types of timespaces interwoven in our two case study firms post-2000 and the implications for the demand for travel. Though many business practices involve the interweaving of timespaces, given our research focus, our discussion is confined to coordinated actions involved in the procuring and arranging of work.

12.2   Historical Changes to the Interweaving of Timespaces

The market for construction and engineering consulting services is related to the growth of industries and construction, the ebbs and flows in public and private spending on infrastructure and manufacturing, and the deskilling within local and national government in the move towards a prevalence of consulting (Kakabadse and Kakabadse 2001). Historically, in the case of our two study firms, procuring and arranging work was focused at the regional level, as defined by a particular firm office. For example, in procuring work, profit and loss accounts for the firms ensured project teams were drawn from offices within a particular national region, and often only one office. This arrangement, which was encouraged and facilitated by common and even harmonised interwoven timespaces, ensured that the coordination benefits that ensued were relatively easy to achieve. Interpersonal structuring, made possible by permanent or easily facilitated co-presence, was the dominant way of interweaving timespaces. Consultancies conducted their work within one office and/or on the project site, or made occasional physical trips to the client based on the expectation of personal interaction between the provider and the client (Baark 1999). Demand for travel to interweave timespaces, whilst not absent, was therefore low.

The relatively constrained objective spatial and temporal qualities of the procurement and arrangement of work came to be unsettled by a number of events; notable being globalisation and structural changes to the engineering consulting industry, computer software and networks, international regulatory reforms and airline de-regulation. These changes led to a turn towards proactive global 'market seeking' strategies. Engineering consultancies began to redefine "themselves in terms of their entrepreneurial endeavours" (Coyne et al. 1996: 749) and open new offices or divisions abroad. Establishing an office was initially driven by the 'follow the client' logic. Opening offices outside the UK continued to be staffed with consultants able to deliver local projects for the client in question, thus continuing to rely on timespaces interweaved through interpersonal structuring. However, a number of changes quickly undermined this state of affairs.

In the 1980s, personal computer workstations became ubiquitous and gradually indispensable in the operation of engineering consultancy firms. Initially, the financial costs of linking computer workstations with Computer Aided Design (CAD) software and other emerging telecommunication networks was financially prohibitive except for within large firms. These telecommunication systems eventually spread throughout the consulting sector, becoming taken for granted and encouraging the rise of global divisions and 24-hour working patterns (Tombesi 2001). Consequently, whilst historically project team members (designers, engineers, architects) would meet daily around the drafting table to accomplish project outcomes, by the 2000s this had changed significantly. Although drawings remain a critical tool for shared interpretation, the introduction of CAD and ICT into the arrangement of work resulted in not only fewer routine technical workers (Coyne et al. 1996; Baark 1999) but also a greater tendency to work in teams that share documents electronically. Such changes began to blur the boundaries of working time and location (Vendramin et al. 2000) as project teams became spatially stretched and temporally dispersed—something new accounting systems and performance assessments further encouraged (Buch and Andersen 2015). Thus, both client-facing and internal-firm features of our case study firms began to emphasise communication skills and production rhythms conducive to coordinating actions where objective space was stretched and time dispersed (different place, same time; different place, different time). This emphasis alludes to how chains of action can re-bundle various business practices (see Schatzki 2012b: 76–78). Chains of actions described below which constitute how coordinated actions involved in the procuring and arranging of work occur today (in the future), linked through past chains of action discussed here, came to encourage the interweaving of timespaces where objective space and time were less constraining features of coordinating action.

Concurrently, the de-regulation of the airline industry, beginning in the 1970s but having noticeable effects in the 1990s and later, lessened the financial burden of flying experts around the country or the world. This change does reflect a simple supply creates demand scenario. Yet, de-regulation also further normalised time-space compression and the wider global trend towards globalised work. For example, in our two case study firms, ease of flying strengthened the geographical reach of each firm. In turn, firm strategies began to target new business opportunities both within the UK and abroad. Firm strategies began to stipulate and normalise the procuring of work based on specific projects and skill sets within the wider firm, not within any one particular office. This change rendered travel-enabled work practices not only cheaper but also indispensable. New firm procurement strategies, therefore, supported and encouraged a re-organisation of teams and offices, as senior staff and project managers began to observe that a single office no longer needed to contain a full array of professional skill sets, as potential workers could be drawn from different offices.

Spatially stretched and temporally dispersed actions had significant implications for the interweaving of timespaces. The emergence of orchestrated timespaces in which different aims are pursued (e.g. two offices chasing very different types of work) despite interdependence (all offices needing to procure similar work that can be shared globally) risked undermining coordination. No longer was interweaving achieved solely by interpersonal structuring. Chains of action, brought about by changes in technology and airline deregulation, ensured that work passed between offices brought a constellation of actors into interdependence. Further, new intentional directedness emerged as actors in different places but doing similar work came to be directed towards one another, for instance through membership of the same global practice group.

In summary, over a couple of decades, timespaces for the engineering consulting business sector, as illustrated by our two case study firms, shifted from predominantly common timespaces with local characteristics (same place, same time; same place, different time) towards spatially stretched orchestrated timespaces (different place, same time; different place, different time). This implies that over time, in order to ensure coordination of actions related to procuring and arranging of work, more elaborate corporate tactics became necessary for interweaving timespaces. The rise in demand for travel was intimately related to this adoption of new tactics.

12.3   How Timespaces Are Interwoven Today

Today, each of our two case study firms have grown in size and broadened their capabilities so that they are capable of acting as a 'one stop shop' (UNESCO 2010; Buch and Andersen 2015). This began in the late 1980s as the sector moved towards design-build coalitions as software and communication innovations called for the increased integration of diverse disciplinary knowledge and skills. Specialising in large projects such as airport construction, for example, allows a firm to sell and deploy their full array of specific skills sets, from geotechnical, seismic and acoustics, to civil and structural engineering and architecture, irrespective of objective space and time constraints which are now routinely overcome through various changes discussed above.

Whereas our two case study firms were historically organised regionally within countries, today, project teams are structured globally by discipline, often irrespective of office location. In addition, firm profit and loss accounts are no longer based on office, but on disciplines and professional expertise. Further, intranet capabilities in our two firms provide employees with databases to share ideas for procuring work, client information, bid information, and so on. These arrangements involve virtual ways of working, and thus a mix of very complex common, shared and orchestrated timespaces suitable to overcome the constraints of objective time and space.

As Schatzki's (2012b: 52) perspective reminds us, the timespaces of any one actor for any given practice is "partly common, partly shared, and partly personal". In the case of our firms, "organisations circumscribe the shared dimension of their participants' timespaces because teleoaffective structures comprise acceptable futures in addition to enjoined ones" (Schatzki 2012b: 53). The importance of acceptable futures can be observed in how, today, practice groups and teams operate in our case study firms. All individuals are members of a global practice group and are expected to contribute to the strengthening of that group's competitiveness. Meanwhile, work is organised in teams which are staffed by the most qualified individual regardless of location, subject to availability at the required time. In both, there is a risk of orchestrated timespaces undermining the strategic priorities of the firm, given that in these, coordination is difficult to achieve. Common timespaces and enjoined actors, and ideally harmonised timespaces in which there is "seamless interlocking" (Schatzki 2012b: 88), are needed. As such, both firms invest in developing and nurturing trust and respect and a sense of shared agenda, as both are central to ensuring good professional relationships and ultimately movement towards enjoinment and harmony.

The complex arrangements required to ensure coordination in our two firms today can be illustrated through the following examples. A CEO recently wrote to all staff to indicate that being more geographically mobile was an increasingly important part of business as usual for the firm. From our perspective, the letter further embedded existing conventions and norms around working for a global-engineering consulting firm. Specifically, the letter underscores the centrality of travel in the production of shared and harmonised timespaces associated with procuring and arranging work. For existing staff there was a degree of reluctance (e.g. due to family arrangements) but nonetheless this was seen as part of new terms and conditions for graduate jobs and acknowledged as a way of getting on in the firm. Indeed, Key Performance Indicators at senior management level included being visible, mobile and geographically agile. This reflects an acceptance that whilst virtual communications are acceptable in some cases (principally non-client communication), and increasingly a part of some working practices (e.g. CAD production in India), travel and the creation of common and harmonised timespaces that it allows is unavoidable. Hence demand for business travel in our case firms is a result of its critical role in interweaving timespaces and ensuring coordination when work is procured and completed.

12.4 Conclusion

This chapter has provided an account of how demand for business travel is intrinsically related to how business actions are organised spatially and temporally. Schatzki's (2009, 2012a, b) notion of timespace highlights how our case study firms constitute 'a bundle of practices and material arrangements' (Schatzki 2005, 2006a). Such features did not emerge overnight, but reflect several decades of changes which have occurred within the wider professional services sector. Over time, the interweaving of timespaces, which is essential to coordinate actions related to procuring and arranging of work, has become more elaborate as tasks, activities and decisions have been stretched across different places and dispersed in time. Simply said, procuring work back in 1920 is different to how work is procured today. This history illustrates the broad, fluid and historical preconditions of the sites as well as the material arrangements that prefigure how our firms bid for and arrange work. With respect to demand for travel, changes to the interweaving of timespaces mean that travel, as in broader society, has become more inscribed into the profession of engineering consultancy and in the firms in our case study.

As organisations help to circumscribe the context for coordinating actions and thus help to interweave timespaces, the concept of timespaces offers scholars an alternative way to think about how past organisation practices shape current work practices, how work practices continue to evolve, and how work practices might be steered to promote temporalities and spatialities that reduce demand for travel. As other chapters in this book also reflect on, drawing attention to the spatialities and temporalities of energy practices helps to make sense of the historical, current and future variability and transformations of energy demand. Important spatial and temporal features which demand business travel can be observed in a variety of ways. For instance, changes in firm strategies and organisational restructuring, affordances brought about through ICT (e.g. telecommunications, CAD, inter/intranet, servers, virtual communication) and mobility (e.g. low cost airlines and regional airports, travel management companies, company cars, high speed trains) helped to spatially stretch and temporally disperse actions. Thus, objective space and time came to be less and less constraining. This relates to changes in how 'lives hang together' and interwoven timespaces are developed (Schatzki 2012b). Given the tendency for spatial stretching (actions at different places at the same or different times), interpersonal structuring of timespaces is now more difficult and does not happen organically. As a result, the interweaving of timespaces through chains of action and intentional directedness becomes important, but both require greater seeding and effort if coordination is to be ensured. Therefore, over time, our two firms have evolved to embed mobility in how each bids for and arranges work. Hence, neither firm could operate without the travel that for environmental and other concerns they may wish to reduce.

Our findings and focus on changes in the interweaving of timespaces strongly suggest that questions about if and how technology (video conferencing in particular) might substitute for travel need to be complemented by perspectives on what creates demand for travel in the first place. If, as research suggests (Strengers 2015; Haynes 2010), technology cannot substitute for the well-known affordances of corporeal co-presence, then understanding how demand for travel is made becomes essential so that multiple ways of unmaking it become possible. Space-time compression brought about by changes in technology is central to understanding today's global economy in which our firms operate. However, Schatzki (2012b: 87) notes that the transformation and redesign of interwoven timespaces were also central to space-time compression: "changes in social life require changes in temporalspatiality." As such, to reduce travel means considering different ways to interweave timespaces.

An obvious answer is to reduce the spatial stretching and temporal dispersion that has meant travel is needed to allow chains of action and intentional directedness, and in turn coordination. Yet, even in the current moment when there seems to be growing scepticism as far as globalisation is concerned, a return to localised business is hard to imagine. The strategic locating of different elements of the bidding and work delivery process might offer one compromise (Jones 2013). Given the insights gained into the need for travel to interweave timespaces and ensure coordination, it might be possible to identify the individuals and practices that require most travel and to co-locate these whilst maintaining global operations and clients. Alternatively, the question of technological substitution could be reimagined. Instead of asking how technology can reproduce the affordances of embodied encounter, research might ask how the interweaving of timespaces can be achieved via technology. This is a subtly different question because it highlights the possibility that coordination achieved as a result of chains of action and intentional directedness might be enabled by technology. Recall that coordinated action is only possible when timespaces are interwoven (Schatzki 2012b: 87), and that interweaving can occur in the: same place, same time; different place, same time; different place, different time; or same place, different time. As such, coordinated actions might be enabled and associated by technologically facilitated following/reacting to others (for chains of action) and/or focus on the actions/emotions of others (intentional directedness) that is not simply about reproducing meetings via video conferencing or alike. For instance, might technology generate forms of continual connection in ways that allow the development an awareness of and ability to respond to what others do in different times and places? This might involve forms of visualisation that already exist in project management and intranet tools, but enhanced. There is much to work through in relation to such an idea, but the key point is that perhaps we need to imagine new ways of interweaving timespaces that are not limited only to travel or the technological substitution of meetings. Technologically interweaved timespaces might, then, be a productive agenda for research.

Acknowledgements This work was supported by the Engineering and Physical Sciences Research Council [grant number EP/K011723/1] as part of the RCUK Energy Programme and by EDF as part of the R&D ECLEER Programme.

Notes

1. Though Schatzki uses the term activity timespace to differentiate it from Heidegger's notion of timespace which is also a crucial feature of human life, our use of timespace accords with the concept as developed by Schatzki.

Bibliography

Aguilera, A. 2008. Business travel and mobile workers. *Transportation Research Part A: Policy and Practice* 42: 1109–1116.

Anable, J., A. Darnton, K. Pangbourne, et al. 2015. *Evidence base review of business travel behaviour to inform development of a segmentation of businesses.* Main report, University of Aberdeen (PPRO 04/06/43). Available at https://www.gov.uk/government/uploads/system/uploads/attachment_data/file/415569/business-travel-behaviour-main-report.pdf

Arnfalk, P., and B. Kogg. 2003. Service transformation—Managing a shift from business travel to virtual meetings. *Journal of Cleaner Production* 11: 859–872.

Baark, E. 1999. Engineering consultancy: An assessment of it-enabled international delivery of services. *Technology Analysis & Strategic Management* 11: 55–74.

Bryman, A. 2008. *Social research methods.* Oxford: Oxford University Press.

Buch, A., and V. Andersen. 2015. Team and project work in engineering practices. *Nordic Journal of Working Life Studies* 5: 27.

Cohen, M.J. 2010. Destination unknown: Pursuing sustainable mobility in the face of rival societal aspirations. *Research Policy* 39: 459–470.

Coyne, R.D., F. Sudweeks, and D. Haynes. 1996. Who needs the internet? Computer-mediated communication in design firms. *Environment and Planning B: Planning and Design* 23: 749–770.

Davidson, R., and B. Cope. 2003 *Business travel: Conferences, incentive travel, exhibitions, corporate hospitality and corporate travel.* Pearson Education, Financial Times Press.

Dicken, P. 2011. *Global shift: Mapping the changing contours of the world economy*. London: Sage.

Faulconbridge, J.R. 2006. Stretching tacit knowledge beyond a local fix? Global spaces of learning in advertising professional service firms. *Journal of Economic Geography* 6: 517–540.

Faulconbridge, J.R., J.V. Beaverstock, B. Derudder, et al. 2009. Corporate ecologies of business travel in professional service firms working towards a research agenda. *European Urban and Regional Studies* 16: 295–308.

Gallié, E.-P., and R. Guichard. 2005. Do collaboratories mean the end of face-to-face interactions? An evidence from the ISEE project. *Economics of Innovation and New Technology* 14: 517–532.

Gustafson, P. 2006. Work-related travel, gender and family obligations. *Work, Employment and Society* 20: 513–530.

———. 2012. Managing business travel: Developments and dilemmas in corporate travel management. *Tourism Management* 33: 276–284.

Harvey, D. 1989. *The condition of postmodernity: An enquiry into the origins of cultural change*. Oxford/Cambridge, MA: Blackwell.

Haynes, P. 2010. Information and communication technology and international business travel: Mobility allies? *Mobilities* 5: 547–564.

Hydle, K.M. 2015. Temporal and spatial dimensions of strategizing. *Organization Studies* 36: 643–663.

Jones, A. 2007. More than 'managing across borders?' The complex role of face-to-face interaction in globalizing law firms. *Journal of Economic Geography* 7: 223–246.

———. 2013. Conceptualising business mobilities: Towards an analytical framework. *Research in Transportation Business & Management* 9: 58–66.

Kakabadse, A., and N. Kakabadse. 2001. Outsourcing in the public services: A comparative analysis of practice, capability and impact. *Public Administration and Development* 21: 401–413.

Lassen, C. 2010. Environmentalist in business class: An analysis of air travel and environmental attitude. *Transport Reviews* 30: 733–751.

Lyons, G. 2013. Business travel—The social practices surrounding meetings. *Research in Transportation Business & Management* 9: 50–57.

Millar, J., and J. Salt. 2008. Portfolios of mobility: The movement of expertise in transnational corporations in two sectors—aerospace and extractive industries. *Global Networks* 8: 25–50.

Neuman, W.L. 2011. *Social research methods: Qualitative and quantitative approaches*. Boston: Pearson.

Orr, J.E. 1996. *Talking about machines: An ethnography of a modern job*. Ithaca: Cornell University Press.

Reckwitz, A. 2002. Toward a theory of social practices: A development in culturalist theorizing. *European Journal of Social Theory* 5: 243–263.

Salt, J., and P. Wood. 2012. Recession and international corporate mobility. *Global Networks* 12: 425–445.

Schatzki, T.R. 2005. Peripheral vision: The sites of organizations. *Organization Studies* 26: 465–484.

———. 2006a. On organizations as they happen. *Organization Studies* 27: 1863–1873.

———. 2006b. The time of activity. *Continental Philosophy Review* 39: 155–182.

———. 2009. Timespace and the organization of social life. In *Time, consumption and everyday life: Practice, materiality and culture*, ed. E. Shove, F. Trentmann, and R. Wilk, 35–48. Oxford: Berg.

———. 2012a. A primer on practices. In *Practice-based education: Perspectives and strategies*, ed. J. Higgs, R. Barnett, S. Billett, et al., 13–26. Rotterdam: Sense.

———. 2012b. *The timespace of human activity: On performance, society, and history as indeterminate teleological events*. Lanham: Lexington Books.

Sheller, M., and J. Urry. 2006. The new mobilities paradigm. *Environment and Planning A* 38: 207–226.

Shove, E., M. Pantzar, and M.T. Watson. 2012. *The dynamics of social practice: Everyday life and how it changes*. London: Sage.

Shove, Elizabeth, G. Walker, et al. 2014. What is energy for? Social practice and energy demand. *Theory, Culture & Society* 31: 41–58.

Storme, T., J. Faulconbridge, J.V. Beaverstock, et al. 2016. Mobility and professional networks in academia: An exploration of the obligations of presence. *Mobilities*: 1–20.

Strengers, Y. 2015. Meeting in the global workplace: Air travel, telepresence and the body. *Mobilities* 10: 592–608.

Thrift, N.J. 2008. *Non-representational theory: Space, politics, affect*. London: Routledge.

Tombesi, P. 2001. A true south for design? The new international division of labour in architecture. *ARQ: Architectural Research Quarterly* 5: 171–180.

UNESCO. 2010. *Engineering: Issues, challenges and opportunities for development*. United nations educational, scientific and cultural organisation, Paris. Available at http://unesdoc.unesco.org/images/0018/001897/189753e.pdf

Urry, J. 2003. Social networks, travel and talk. *The British Journal of Sociology* 54: 155–175.

———. 2012. Do mobile lives have a future? *Tijdschrift voor Economische en Sociale Geografie* 103: 566–576.

Vendramin, P., G. Valenduc, I. Rolland, et al. 2000. *Flexible work practices and communication technology.* Report for the European Commission, SOE1-CT97–1064, European Commission, DG Research. Downloaded on 13: 2007.

Wickham, J., and A. Vecchi. 2009. The importance of business travel for industrial clusters–Making sense of nomadic workers. *Geografiska Annaler: Series B, Human Geography* 91: 245–255.

World Travel and Tourism Council. 2011. *Business travel: A catalyst for economic performance.* London: World Travel and Tourism Council.

Yanow, D., and P. Schwartz-Shea. 2006. *Interpretation and method: Empirical research methods and the interpretive turn.* New York: Sharpe.

Ian Jones is a Research Fellow in the Institute for Transport Studies at the University of Leeds, UK. His research interests relate to professional urban and transport practices and the (re)allocation of urban road space with respect to sustainability, mobility and mobility systems. Recent research completed as part of the DEMAND Centre focuses on mobility related to business travel and the future of online shopping.

James Faulconbridge is Professor of Transnational Management and Head of the Department of Organisation Work and Technology at Lancaster University, UK. His research examines a range of issues relating to the globalisation of professional service firms, with particular focus upon the way knowledges and practices are reproduced and transformed within firms as they move across space. Recent research completed as part of the DEMAND Centre concerns how building design professionals develop knowledge about what is legitimate and normal in relation to building design, and how business travel has become an expected, normal and needed form of mobility.

Greg Marsden is Professor of Transport Governance in the Institute for Transport Studies at the University of Leeds, UK. His research interests relate to the why and how of policy making and in particular the interaction between different agents and agencies in the policy process. He works extensively on issues surrounding climate change, resilience and energy in the transport sector (and beyond). Recent research completed as part of the DEMAND Centre concerns mobility and understanding whether we can design solutions which improve well-being but do not inherently require greater mobility.

Jillian Anable is Professor of Transport and Energy in the Institute for Transport Studies at the University of Leeds, UK. Her research focuses on transport and climate change with particular emphasis on the potential for demand-side solutions to reduce carbon and energy from transport. Recent research completed as part of the DEMAND Centre concerns attitudes to transport, energy and climate change with respect to trends and patterns in energy demand, business travel and energy-related economic stress in the UK.

Part 5

Shifting Rhythms

The hotel air conditioning whirred and hummed, disturbing my sleep. It seemed to have a mind of its own. I had pushed some buttons earlier when I arrived from the airport at 2 in the morning, but nothing much seemed to happen. Nearly time to get up. The sun was starting to come through the curtains, the clock turned over, a newspaper dropped outside my door. The smell of breakfast drifting through the corridor, other people stirring, TVs coming on, showering and washing. Another day beginning. Another same old, same old for the hotel.

The temporal orderings and patterns that make up the everyday, and that give rhythm both to how the social world rolls out and to how energy demand fluxes and flows, are at the core of the two chapters in this final part of the book. Both make clear that while temporal ordering is to some degree shared across societies, in terms of for example broadly shared weekly or seasonal rhythms, they are also grounded within particular settings and situations—those of hospitals and hotels in the cases these two chapters examine, but we could imagine many others. Rhythms of practices and those performing them interweave and interact in such settings. They are shaped by the ways that different activities interrelate and bundle together, and by the expectations and standards of performance that are intrinsic to performing well, for example as a hospital or as a hotel. The combined beats, pulses and patterns that are generated do not though just stand still, in stasis. So we can also have an interest in understanding something about the movement of rhythms in time, both in terms of how they shift and transform and how they might be purposefully shifted around to particular ends. Such a phenomenon is increasingly sought after in the energy world. As explained in the introduction to the book, both peaks in demand that are putting pressure on the electricity grid and troughs in demand that are not using the available renewable energy supply on a very sunny or windy day, mean that forms of demand flexibility and responsiveness are increasingly aspired to by energy policy makers and grid managers. There is now money to be saved (or to be made), and carbon to be cut, by flexing energy use in response to such temporally structured energy system objectives. Both hospitals and hotels are settings where 'demand responsiveness' is being imagined as possible. But how much purposeful flexing to extant, continually reproduced rhythms of energy use can really be achieved, and what bounds and limits these possibilities, is another question.

While we can approach these two chapters through a common lens of rhythm and flexibility, in other ways they are very different, maybe more so in fact than any other two chapters in this book. One is strongly sociological engaged in theoretical development and application; the other is more pragmatically grounded in, for example, details of the characteristics of energy-using technologies and their 'load profiles'. One begins with the applied objectives of turning off and on energy-using technologies at particular times, tracking the actual performance of hotel-located technologies doing just that; the other addresses such possibilities only as an implication of the analysis of interrelated institutional and working rhythms that is the main theoretical concern, with the hospital an intriguing and productive empirical example. Such are the many different approaches through which a concern for investigating demanding energy can be pursued.

For Mitchell Curtis and colleagues then the hotel is primarily framed as a setting for energy-using technologies—lights, heating and ventilation systems, kitchen equipment—each of which can be characterised both in relation to how fast they can be powered down and back up again (which they term responsiveness) and how feasible it is in practice to do so in response to 'signals' which come from the energy supply system (which they term flexibility). Responsiveness is to some degree a part of the temporal features of the technology itself, whereas they are clear that flexibility is a matter of what the energy-using technology is *for* in the hotel context—what service is being provided, how this relates to expected standards of customer experience, what temporal pattern this service needs to be available over and what the implications of 'powering down' are likely to be for the quality of service provision. The rhythms of 'doing business', of doing it well and of working with the temporal patterns and expectations of customers arriving, leaving, eating, sleeping and so on, strongly condition and differentiate the possibilities of turning things on and off. The rhythmic demands of the institutional setting, as well as the energy-using technologies that inhabit it, are vital to any realistic assessment of how much scope there is for demand flexibility.

Stanley Blue starts in a different place, but would have no problem agreeing that the situated institutional setting is important to investigate and understand. He conceptualises hospital life as being full of interconnected practices and as underpinned by a socio-temporal structure that shapes 'normal' and acceptable ways of working. The timing of a given practice is, he argues, a product of its connections to other practices—so testing comes before diagnosis, before treatment and so on—as well as of the temporal ordering of the hospital as a whole, and also of particular material arrangements and the configuration of professional and jurisdictional boundaries. All of these interconnections, he emphasises, need to be understood in terms of their historical formation and development. Hospitals, clearly, are complex, evolving and multifaceted settings for the making of energy demand. To find a way through this potentially overwhelming web of interrelations, he takes particular examples—such as changes to the organisation of breast cancer screening services—to show how there is sometimes flexibility and sometimes fixity in interconnections, and that these in turn present opportunities and obstacles to forms of temporal reorganisation. It follows in relation to energy demand that intervening in the organisation and timing of working arrangements is in principle an opportunity for achieving 'demand response' and reducing demand overall; but realising this depends on there being sufficient flexibility in the various forms of connection that are holding the temporal arrangements of hospital life together.

Both chapters then provide much for others interested in the rhythms of energy demand to build upon, both within and beyond the institutional settings that they have selected as their empirical focus.

13

Demand Side Flexibility and Responsiveness: Moving Demand in Time Through Technology

Mitchell Curtis, Jacopo Torriti, and Stefan Thor Smith

13.1 Introduction

Demand Side Response (DSR) consists of a set of programmes, policies and technologies that enable shifting of electricity demand in time with varying degrees of end user's engagement. It is increasingly seen as the main technique for providing flexibility of demand to energy systems, which until recently have largely worked on the basis that demand is given, with little scope for addressing the intensity of peak loads or of matching the timing of demand to available fluctuating supply. It is probably unsurprising that the literature on DSR features mainly analyses of its techno-economic impacts. Most of the existing studies tend to focus either on the need for DSR at different levels, that is national balancing, distribution networks, and so on (Strbac 2008; Bradley et al. 2013) or on the technical features of DSR (Siano 2014; Hong et al. 2012) and its

M. Curtis (✉)
Technologies for Sustainable Built Environments Centre, University of Reading, Reading, UK

J. Torriti • S. Smith
School of the Built Environment University of Reading, Reading, UK

© The Author(s) 2018
A. Hui et al. (eds.), *Demanding Energy*, DOI 10.1007/978-3-319-61991-0_13

alternatives for the provision of flexibility (Poudineh and Jamasb 2014). Little emphasis has been placed on the practical steps for DSR implementation and the specific loads contributing to its delivery, or on the extent to which flexibility can be delivered through automation. In some instances, techno-economic research has attempted to shed light on what a flexible load is. For instance, Silva et al. (2011) define the flexibility of appliances as depending on their operation patterns and uses. Some appliances, such as refrigerators and freezers, have a nearly constant electricity demand. Others, such as heat pumps and chillers used for water heating and space cooling/heating, are used periodically/intermittently throughout the day. Washing machines and dishwashers consume electricity during a fixed duration cycle. However, such categorisations are limited by how flexibility in demand is being understood and defined, and also by how different patterns of responsiveness—the speed at which a demand reduction can be achieved—are being presumed or seen as integral.

Our main objective in this chapter is to move beyond existing approaches to better incorporate the material technological arrangements of appliances and infrastructures and the social rhythms of everyday coordination into analysis of DSR in practice, and in so doing to propose clearer definitions of how flexibility and responsiveness should be understood. Taking the example of hotels as a site of energy demand, we detail which energy loads have potential for demand responsiveness and focus on questions of automation at different stages of the DSR process. The following questions are addressed: (i) what is a flexible load from an energy service or appliance point of view; (ii) what is a responsive load from an energy service or appliance point of view; and (iii) to what extent can DSR take place without human intervention?

We begin by reviewing established approaches and their limitations before laying out the definitions of flexibility and responsiveness that are utilised across the rest of the chapter. We then identify the typical appliance and service loads of a hotel, describe flexible and responsive loads in hotels, distinguish between stages in which manual and automated DSR take place, and finish by discussing the findings and wider implications of this work.

13.2   Existing Approaches to Flexibility and Responsiveness and Their Limitations

Social science research has taken different directions in the attempt to define flexibility in demand and how this could be achieved. Accounts of flexibility have been dominated by approaches centred on behavioural science and economics. Examples abound, but two are presented here as they epitomise the prevalence of concepts of 'direct feedback' and 'price elasticity' in explaining how flexibility might take place.

The first example is the European Commission's provision in the 2012 Third Energy Package which set a target of 80% roll-out of smart meters by 2020. One of the key principles underpinning this unprecedented large-scale European project is that direct feedback of information to the consumer (domestic or non-domestic) can increase demand flexibility. The effects of direct feedback are based on psychological theory of feedback, which dates back to the 1930s, with Skinner's (1938) model of operant conditioning: this states that behaviours which produce a positive effect are more likely to be repeated than those which produce a negative effect. According to this view, positive results could be seen as a positive reinforcement and negative results as a punishment, thus respectively encouraging or discouraging subsequent behaviour. In the 1960s, feedback was connected with goals and information about progress towards goals (Bandura 1969). In the 1990s Feedback Intervention Theory appeared and asserted that behaviour is regulated by comparisons to pre-existing goals, as behaviour is generally goal directed, and that people use feedback to evaluate their behaviour in relation to their goals. Both negative and positive feedback, it is claimed, can bring about behavioural change (Vallacher and Wegner 1987). These theories suggest that feedback is effective provided that the individual focuses on the feedback goal. This means that feedback will have to direct the attention of the individual to a specific goal compared with the status quo. Availability and visibility feedback are two necessary conditions for the success of feedback.

At an empirical level, over a 100 empirical studies of feedback of energy consumption information have been conducted over the past 40 years and over 200 articles have been published about energy feedback during that time (Karlin et al. 2014). Studies specifically concerned with newly installed smart meters and the flexibility associated with the real-time information flows they can provide have generated mixed findings, but most widely cited fact is that they can lead to between a 3 % and 15 % reduction in peak consumption. However, some of the more optimistic findings have been based on limited research designs whereas recent advances in smart grid technology have enabled larger sample sizes and more representative sample selection and recruitment methods for smart metering trials. Trials from the UK Energy Demand Research Project show reductions of peak demand of around 6–10% from direct feedback (Ofgem 2017).

A second example of an approach to understanding flexibility, from the UK, concerns an initiative to introduce dynamic tariffs following recommendations by the Competition and Markets Authority (2016). Dynamic tariffs are designed to be cost-reflective and include time of use tariffs, which encourage consumers to shift their consumption away from times when demand on the network is higher (peak periods) or reduce it altogether. In other words, dynamic pricing should increase the flexibility of the demand side thanks to the exposure to a varying price of electricity. The main principle supporting the effectiveness of dynamic tariffs is price elasticity, that is the willingness and ability of individual households to respond to price signals by altering energy consumption. Price elasticity estimates for residential electricity demand vary widely across the literature. Alberini and Filippini (2011) review a number of studies and suggest that differences might be due to the sample period, the nature of the data, geography and level of aggregation of the data. Recent empirical studies show that the price elasticity of electricity consumption ranges from as low as −0.06 (Blázquez et al. 2013) to as high as −1.25 (Krishnamurthy and Kriström 2013).

In general, literature reviews in this area seem to suggest that the price elasticity of demand for electricity is low. For instance, a meta-analysis by Espey and Espey (2004) finds that the median short-run elasticity for 36 studies is −0.28. So, for intra-day price elasticity (or extreme short-run price elasticity) there seems to be evidence that parents, for example, do not postpone by an hour the moment when they drop children at school to take advantage of a profitable off-peak tariff. A recent study by a group of Swedish researchers looks at elasticity changes during the day, which seem to be higher for lighting and cooking than for heating (Broberg et al. 2015). Hence, the engagement of end users in dynamic tariffs depends mainly on the extent to which everyday life and household routines can be reconciled with the new tariffs (Buryk et al. 2015). Different loads therefore play an important role in price elasticity of electricity demand.

In reviewing these two examples, based on behavioural and economic approaches, we can see two sets of definitions of flexibility and responsiveness being deployed. Behavioural approaches measure flexibility in terms of the extent to which people respond to direct feedback and responsiveness as the (seldom measured) time it takes to react to real-time information on load profiles. Economic price elasticity approaches define flexibility in terms of the extent to which consumers respond to short-term variations in price and responsiveness is assumed (but seldom measured) as the time it takes to respond to notified price variations.

13.3   Re-defining Flexibility and Responsiveness

Approaches based on direct feedback and elasticity point to causal interpretations of why flexibility and responsiveness occur (Lutzenhiser 1993), but do not take account of the material arrangements involved (including technical characteristics of loads), or the everyday social settings and temporal rhythms in which energy demand is embedded. Interventions through social and technological agents (such as DSR) aimed at increasing the flexibility of energy demand need to be analysed in the context in which they are implemented, rather than in more abstract terms, in order to capture the social character of energy use and how it is affected by the material environment. This is becoming recognised in institutional efforts to make the most of social sciences and their interpretation of energy demand. For example, the Department of Energy & Climate Change (DECC) states that 'for an intervention to have any impact, organisations need to either invest in new technologies and materials or promote changes in practices' (DECC 2014: 24).

Better defining what flexibility and responsiveness are is critical for advancing research work on temporality and DSR. According to Nicholls and Strengers (2015), flexibility is concerned with how rhythms change in time and space. Much of the work on flexibility and rhythms is intertwined with peak demand, which is the main intervention focus of DSR. In essence, more flexibility is supposed to deliver reductions in peak demand. For this reason, attempts to describe flexibility have been connected to issues of synchronicity of practices (Walker 2014) and sequencing of practices (Mattioli and Anable 2017). Torriti et al. (2015) operationalise flexibility into an index which varies depending on measurements of the number of activities carried out throughout the day, synchronisation with everyone else, time spent on one's own, spatial mobility and active occupancy. Making space for flexibility involves a similar process to designing intermittency and irregularity into energy systems and infrastructures (Shove and Chappells 2001). Moving away from causal interpretations which place behaviour and elasticity at the centre of energy choices and preferences involves prioritising temporality—in the measurements of flexibility and responsiveness—and investigating arrangements which facilitate provisions of flexibility also in relation to automated DSR interventions.

We therefore suggest investigating flexibility and responsiveness in terms of changes in rhythms of types of demand loads. The definition of flexibility we will use in the rest of this chapter is as follows: *a measure of how rhythms of demand change in time.* This can be operationalised as a technical measure of how patterns of the asset/device 'flex' or change in time and will vary based on several factors including: time of day (i.e. when end users can experience turn-off), duration of change (i.e. how long it can be turned off for), intensity of change (i.e. which percentage of turn-off can occur), and recovery time (how long it takes before another turn-off event can occur again).

With regard to responsiveness, the definition we deploy is as follows: *a measure of how fast the asset/device can respond to an event request, given an economic or technical input (such as a peak or price signal).* The speed at which an asset can change (turn off/turn down) will dedicate what DSR programmes it can be used for. Temporality depends to some extent on human actions and interventions. Looking into the level of automation associated with DSR events reveals the extent to which flexibility and responsiveness can take place without intermediaries.

The existing literature has not as yet focused on load characteristics in order to understand the flexibility and responsiveness in a DSR event for an individual site. What is more, little attention has been paid to levels of automation and how these may or may not occur for DSR events. In what follows we therefore take forward such analysis using the specific case of hotels as a site of potentially flexible and responsive energy demand.

13.4   Hotels as Sites of Demand Flexibility and Responsiveness

Hotels are of particular relevance to work on energy demand and temporal shifting, because they are working places in which people perform social practices typically associated with the household. Since DSR has not yet been implemented in the residential sector, hotels provide a useful ground on which to investigate flexibility and responsiveness of household related social practices and also in consideration of the fact that they are equipped with technology for load shifting. For hotel guests, social practices may include sleeping, dressing, washing, watching TV, working and eating, but they are not likely to include cooking, washing dishes, washing clothes and cleaning, which generally in most hotels in the UK are performed by staff and have temporalities concentrated around work shifts of staff. Hotels are also a significant example of end users currently providing actual turndown DSR (i.e. when devices and appliances are turned off temporarily), as opposed to stand-by generation in which, for example, local stand-by diesel generators are switched on during peak periods to reduce demand on the wider grid (Grünewald and Torriti 2013). We are therefore able to examine the details of the loads in hotels which are being 'turned down', and specifically why heating, ventilation and air conditioning (HVAC) are preferred over other loads for DSR in hotels (i.e. why HVAC loads are considered 'flexible').

The data used is based on nine hotels located in the UK as obtained from a DSR aggregator. Aggregators operate by combining small flexible loads from multiple end users into a virtual single load for the National Grid and take responsibility for managing the DSR process. Later in the chapter the aggregators' role in facilitating DSR will be considered in more depth. The studied hotels range in size from 1100 to 2400 square meters, with between 130 and 390 bedrooms. The data covers 15 DSR events between October 2014 and October 2016. For DSR events the hotels' overall electricity demand in kW was recorded every minute with four hotels having additional sub-metering on the main appliances that formed the turndown capacity. All DSR events were remotely activated by the aggregator and lasted between 15 and 120 minutes.

What Are the Typical Appliance and Service Loads of a Hotel?

Figure 13.1 shows the typical electricity demand split of a standard hotel in the UK as identified in the Chartered Institution of Building Services Engineers (CIBSE) Guide F 'Energy efficiency in buildings' (Butcher 2012). Looking at the loads identified, 'lighting' has the highest demand level, which results from the hotel industry practice of leaving common area lights on 24 hours a day and illuminating outside areas at night time. 'HVAC' demand primarily covers electricity used for air conditioning. Hotels typically use centralised systems for air conditioning, operating large (50–1000 kW) chillers to cool water that is then distributed around the building to air handling units that use heat exchangers and fans to deliver cooling to individual rooms (Carrier 2014). In contrast, the 'heating' component of HVAC is primarily provided by direct conversion of fossil fuel energy and, though electrical ground source and air source heat pumps are seeing greater uptake, currently only a limited electricity demand by circulating pumps and air handling units exists during delivery. 'Catering' in a hotel is normally provided by an onsite restaurant and covers the large-scale cooking and refrigeration requirements. The 'other' and 'office equipment' areas account for small power loads from computers, TVs, phone chargers, printers, kettles and so on.

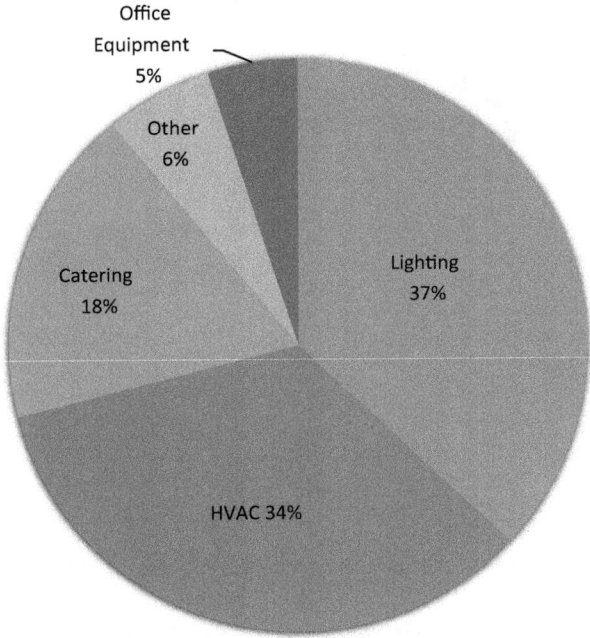

Fig. 13.1   Hotel electricity consumption breakdown (Adapted from Butcher 2012)

Determining the flexibility and responsiveness of each load type requires an understanding of how it is used in the hotel such that potential end user impacts of demand shifting and any technical operating restrictions can be assessed. Figure 13.2 shows a phase space of the responsiveness and flexibility in electrical energy demand of a hotel in schematic form. The schematic helps inform an understanding of the relationship between flexibility and responsiveness for each of the hotel electricity demand categories identified in Fig. 13.1. The categories of usage are spread to cover many possible states, in recognition that different technologies, different system operation and function and different user preferences will influence the level of response and flexibility available. High responsiveness indicates that the appliance/service load is turned off or down rapidly, often instantly or within seconds (e.g. a light), while low responsiveness indicates that the load is reduced gradually due to operating restrictions (e.g. large motors often require a gradual stop to

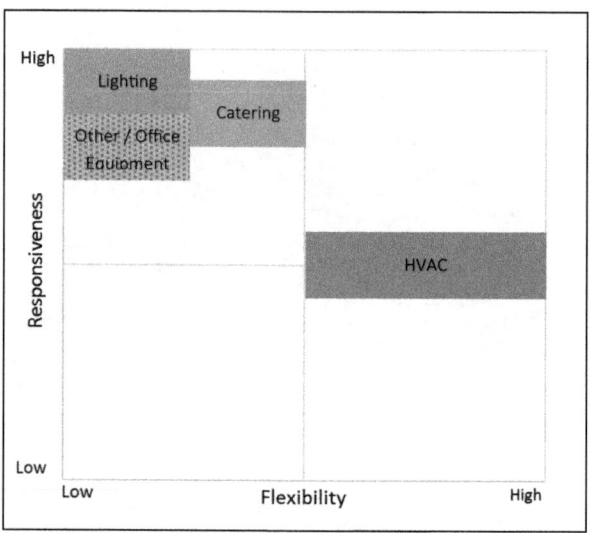

Fig. 13.2 Flexibility versus responsiveness of hotel electricity demand areas

prevent damage). High flexibility indicates that the appliance operation can be temporarily altered with no impacts other than temporary demand reduction (e.g. turning off air conditioning for a short time), while low flexibility indicates there is limited ability to change operation without major impacts (e.g. turning off bathroom lights while in use). Occupying regions of the phase space demonstrates variability in service application and technology operation. To understand the reasoning behind the classifications used in Fig. 13.2, the next two sections of this chapter will focus first on identifying what comprises a flexible load and then on ascertaining what comprises a responsive load for each of the identified hotel electricity demand areas.

13.5   Flexible Loads in Hotels

Based on Fig. 13.2, we consider 'lighting', 'other/office equipment' and 'catering' to have low flexibility, with HVAC being the only demand with high flexibility. Each of these areas will be examined to understand their level of 'flex' starting with the low flexibility areas.

'Lighting' has the highest electricity demand level at 37 % but also the lowest level of flexibility due to end user impacts. As lighting contains no lag in service provision it is difficult to apply demand shifting to occupied rooms where activity and need for lighting cannot be evaluated. If a guest is using a bathroom with no natural light, then the lights cannot be turned off until the room is empty. The installation of presence or movement sensors to enable automated turn-off in corridors and other common areas would represent too greater a cost to warrant installation for DSR participation. Furthermore, the main motivation for having this technology installed comes from energy savings resulting from lights turning off when there is no one present. This in fact negatively impacts on flexibility, as lights need to be on for them to be available to be turned off when demand response is needed.

The range in flexibility offered by 'catering' is a result of function and time of day. During the primary electricity demand periods (breakfast, lunch, dinner) there is low flexibility due to the time dependence on cooking obligations to meet meal requirements for paying guests. However, refrigeration can account for over a third of commercial kitchen electricity demand and offers a potential area for load flexibility (Mudie et al. 2016). The thermal store of refrigerators provides flexibility through the ability to maintain chilled conditions over short periods of time if turned off. Refrigeration also offers a stable and predictable 24-hour demand pattern, no impact to users and short recovery time between DSR events. However, 'catering' as an area is complex to implement with strict constraints set by health and safety for maintaining conditions for food storage, preparation and service. Without using robust and expensive controllers to manage response and ensure that threshold conditions are not exceeded during DSR turndown events, there is little appetite and restricted flexibility available for DSR from 'catering' loads (Grein and Pehnt 2011).

The 'other' and 'office equipment' areas have the lowest electricity demand and represent specific appliances like elevators and pool pumps, or miscellaneous small power loads from many sources including televisions, computers and phone chargers. While offering mixed levels of potential flexibility (e.g. charging could be delayed, yet TV watching probably not), centralising control of the range of devices making up this type of demand is currently technically difficult to implement, yet could become more feasible in the future if appliance makers start including DSR features in devices (e.g. enabling centralised control of charging devices).

'HVAC' systems represent the second largest electricity demand area in a hotel and are deemed as having the highest flexibility. As previously noted, the primary electricity demand in HVAC systems is for the cooling of room temperatures through air conditioning and therefore this will be the context in which HVAC flexibility is reviewed. To understand why this area is categorised as highly flexible requires assessing the following four factors.

First, 'user comfort' is often cited by businesses as a major concern when asked about interest in using air conditioning in demand response programmes (Dolman et al. 2012). Although the literature on comfort and energy demand is vast, there is limited research on comfort in relation to demand shifting. However, ad hoc studies generally indicate that air conditioning can be turned off for approximately an hour before users start to notice (Barton et al. 2013; Xue et al. 2014). Hotels have also started utilising air conditioning in DSR programmes, which implies minimal user impact if it is assumed that a customer-centred service would be very concerned about adopting practices likely to cause guest discomfort (Macalister 2015). The one-hour delay is generally linked to CIBSE's temperature guidance on comfort, which recommends 22 °C with variations of between 16 °C and 28 °C (CIBSE 2015). This variance provides a temperature buffer that enables flexibility through the ability to turn off the air conditioning for short periods of time without users directly noticing. The variance also allows for controls to be put in place that override the demand shift if temperature boundaries are crossed, thereby ensuring minimum impact to users.

Second, one of the reasons air conditioning can be temporarily turned off without obvious impact is due to thermal inertia from two built-in forms of 'storage'. The first occurs due to the thermal properties of buildings, namely thermal mass. The thermal mass in a building will act to delay air temperature increases resulting from internal and external energy gains. The level of inertia is dependent on the building design and construction materials used, as well as gains from people and equipment that go beyond the discussion here. The second form of storage occurs within the air-conditioning system, as these systems for hotels generally work based on centralised cooling of a working fluid (typically water) that is then distributed around a building to be converted into air cooling via air handling units in each room. This means that the air-conditioning system will still provide cooling whilst centralised chiller units are temporarily turned off. This system lag is limited by the demand and available cooling in the working fluid.

Third, the operating schedule of hotel air conditioning offers mixed benefits for flexibility. High levels of flexibility arise during summer through the likely 24-hour operation of the air-conditioning system. However, during winter time the electrical demand for air conditioning drops as heating becomes the dominant temperature requirement, which is met primarily with fossil fuels. While the winter flexibility potential is lower than in summertime, there is still opportunity through turning down the pumps and air handling units.

Fourth, the final element that enables the suitability of air-conditioning systems for load flexibility relies on their controllability and load size. The centralised nature of the main air-conditioning chiller enables easy and inexpensive control, either manually by a local operator or remotely with an addition of a single control system.

13.6   Responsive Loads in Hotels

Determining the responsiveness of an appliance to reduce its electricity load is relatively simple compared to the many factors involved in determining flexibility. Understanding how long it takes to safely turn off or reduce load once a turndown opportunity is identified is the primary consideration for determining appliance responsiveness. One of the main physical features of electricity is that it can be turned off almost instantaneously. However, appliances with some firmware or software control (computers, smart devices) are designed to be turned off in a controlled manner to prevent corruption of data and system operation. Other devices require lagged, sequential ordering of turning off sub-components to avoid deleterious effects to the device's sub-components.

Figure 13.2 shows the hotel electrical demand areas of 'lighting', 'other/office equipment' and 'catering' as being highly responsive due to the general ability to rapidly turn-off appliances within these categories. Within each of these areas there is variance in the responsiveness with 'lighting' having the quickest response, 'catering' being slightly slower due to refrigeration and other large motor-based appliances requiring time to slow down to prevent damage and 'other/office equipment' being the slowest of these areas due to the number of appliances that require a controlled turn-off period (e.g. computers, photocopiers).

In contrast to HVAC's high flexibility, its responsiveness is lower than that of the other hotel demand areas. This is primarily due to main chiller appliances requiring sufficient time to slow down internal pumps to prevent components being damaged from freezing or blockages (sequential, lagged turn-off). Figure 13.3 provides a summary of the response times of nine hotel chiller systems from multiple DSR events. The response time is calculated based on how long in minutes it took between receiving the signal to reduce load and achieving the load reduction. The results show that the overall average response time is 7.5 minutes with a general response time range of 6 to 10 minutes. The high outliers are likely to

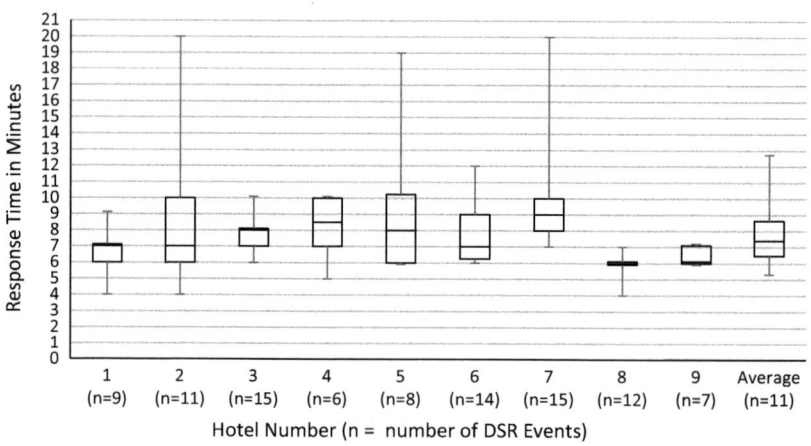

Fig. 13.3   Hotel chiller response times to DSR events

result from communication issues that delayed the chiller responding to the turndown request. The variance shown in Fig. 13.3 demonstrates how the responsiveness of systems to demand shifting is not consistent even for the same equipment and there are clear instances where response has the potential to be much slower than expected from a technical consideration alone. The expectations of responsiveness therefore require careful consideration when deciding on DSR participation, as will be discussed in the next section.

To What Extent Can DSR Take Place Without Human Intervention?

The discussion up to this point makes clear how the profile of flexibility and responsiveness across different categories of demand is dependent on a range of considerations related to the rhythms of different practices that demand energy use: their relation to core business objectives and the customer experience; regulatory requirements on energy service provision; cost implications; and the material capacities and performance of particular technologies. This complex set of considerations also then plays into the degree to which DSR can be entirely automated in any particular setting, with data infrastructures and technologies directly controlling the turning off or down of end use devices rather than manual actions and judgements being involved. By some judgements full automation may be a preferable arrangement, but its realisation is not straightforward.

The ability to provide automated DSR without human intervention depends on the reasons for requiring demand shifting and the parameters of its enablement, including in terms of local flexibility and responsiveness. To explore this subject the hotel example will continue to be used to demonstrate how hotels are enabled for UK DSR services and where automation is currently feasible.

The main driver for DSR services in the UK comes from the main Transmission Operator, the National Grid, who offer programmes that pay for flexible load reductions to help provide cost-effective balancing of the national electricity network (Proffitt 2016). DSR provides a cost-effective balancing method when it is cheaper to pay end users to reduce demand than it is to turn on a large network generator for a short period. The DSR programme used for hotels we studied is called the 'Short Term Operating Reserve' (STOR) and requires providers to deliver a minimum of 3 MW of generation or load reduction which needs to be achieved within 20 minutes of being notified (National Grid 2017a). Hotels, like most end users, seldom have enough load reduction capacity to participate directly, requiring hotels and other small- to medium-sized businesses to use aggregators as an intermediary to participate in DSR programmes. Aggregators operate by combining small flexible loads from multiple end users into a virtual single load for the National Grid and take responsibility for managing the DSR process. Using DSR for providing demand shifting requires interactions between all three parties involved in the process: the National Grid, aggregators and end users. Figure 13.4 illustrates the interaction between each party during this process, including whether steps are manual or automated.

A real DSR event at a hotel in the UK will be used to step through the sequence of events to explain the process flow used in Fig. 13.4. Figure 13.5 shows the timeline of a hotel for 90 minutes in June 2016 when the DSR event occurred. The demand shifting for this event was based on temporarily turning off the main air-conditioning chiller system. The chart in Fig. 13.5 shows the hotel site's overall electrical demand (red line) and the proportion that is used by the chiller (blue line). During this period the following sequence of events occurred:

1. 16:03—The National Gird identifies the need for demand reduction and sends notification to the aggregator that they will need to provide the agreed reductions in 20 minutes.
2. 16:05—Within two minutes the aggregator manually confirms receipt of the request and starts preparation to activate the DSR assets at agreed sites, including the example hotel.
3. 16:08—The aggregator notifies the hotel facilities staff via an automated call that a DSR event will shortly be starting and provides the hotel with the option to cancel participation via selecting 1 on the phone during the automated call.

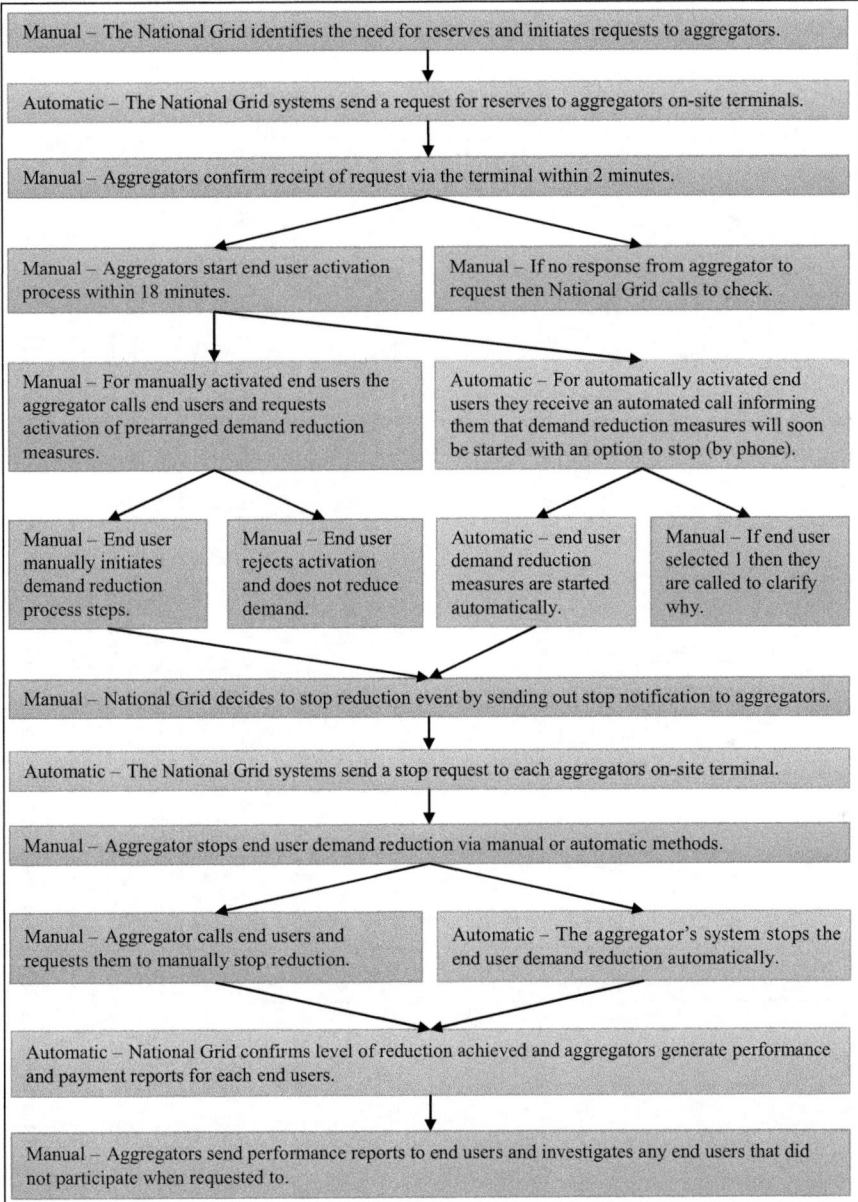

Manual – The National Grid identifies the need for reserves and initiates requests to aggregators.

Automatic – The National Grid systems send a request for reserves to aggregators on-site terminals.

Manual – Aggregators confirm receipt of request via the terminal within 2 minutes.

Manual – Aggregators start end user activation process within 18 minutes.

Manual – If no response from aggregator to request then National Grid calls to check.

Manual – For manually activated end users the aggregator calls end users and requests activation of prearranged demand reduction measures.

Automatic – For automatically activated end users they receive an automated call informing them that demand reduction measures will soon be started with an option to stop (by phone).

Manual – End user manually initiates demand reduction process steps.

Manual – End user rejects activation and does not reduce demand.

Automatic – end user demand reduction measures are started automatically.

Manual – If end user selected 1 then they are called to clarify why.

Manual – National Grid decides to stop reduction event by sending out stop notification to aggregators.

Automatic – The National Grid systems send a stop request to each aggregators on-site terminal.

Manual – Aggregator stops end user demand reduction via manual or automatic methods.

Manual – Aggregator calls end users and requests them to manually stop reduction.

Automatic – The aggregator's system stops the end user demand reduction automatically.

Automatic – National Grid confirms level of reduction achieved and aggregators generate performance and payment reports for each end users.

Manual – Aggregators send performance reports to end users and investigates any end users that did not participate when requested to.

Fig. 13.4 STOR demand reduction manual and automatic process flow steps

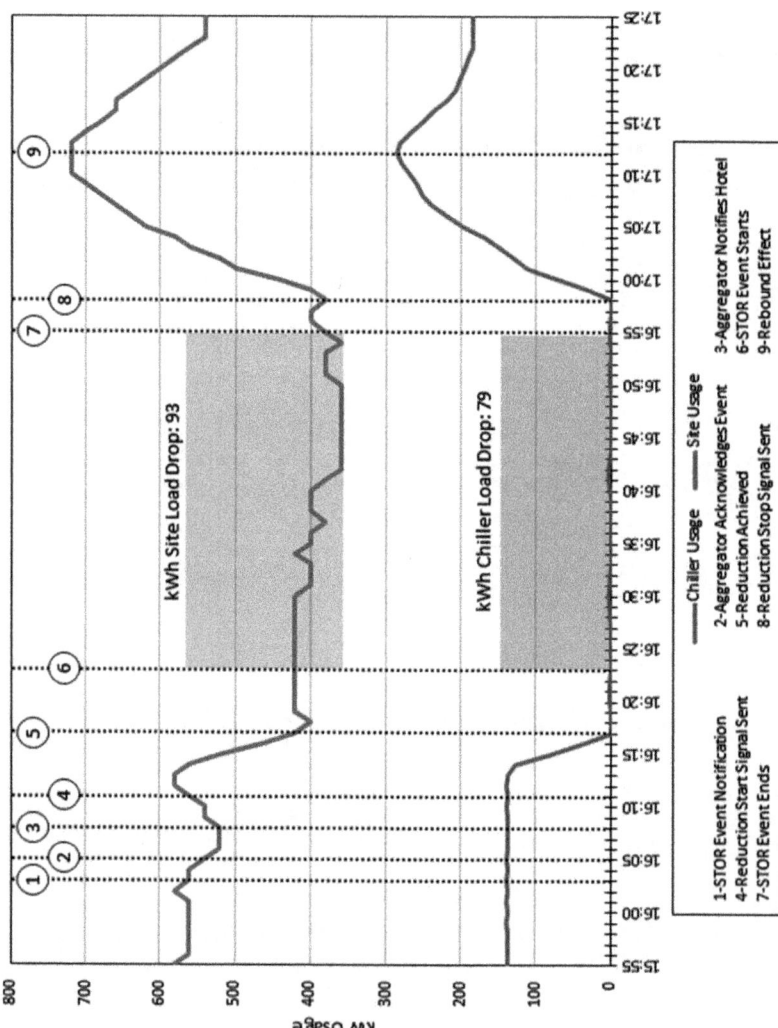

Fig. 13.5 DSR hotel example showing event stages and demand changes

4. 16:11—By answering the call and not selecting the cancel option, the hotel confirms that the hotel is ready to participate in the DSR event. This results in an automatic activation signal being sent to the chiller control system from the aggregator to initiate its demand reduction routine.

5. 16:17—After approximately five minutes the chiller system has effectively turned off with the electricity demand dropping from 144 kW to almost 0 (it still requires a small amount of electricity to maintain system control).

6. 16:23–20 minutes after receiving notification from the National Grid, the event officially starts and is now monitored to ensure the required reduction is achieved.

7. 16:55—A notification is received from the National Grid that the event has finished and that the demand reduction can now stop.

8. 16:57—The aggregator manually triggers the event stop process resulting in a signal being sent to the hotel chillers resulting in them automatically returning to normal demand.

9. 17:12—A rebound effect is noted whereby the chiller system uses more electricity than it would have normally after the event to return the system temperatures back to the pre-event levels.

10. 17:25> − After the event the aggregator will confirm with the National Grid and hotel that they performed as expected by achieving a drop of 79 kWh over the event period. The overall site demand dropped more than this at 93 kWh due to other unrelated electrical demand changes.

This sequence of events demonstrates how in this example a mix of manual and automated steps is required to enable demand shifting. In the example the demand shifting is clearly seen in the lowering of demand during the event and shifting to higher demand afterwards. What can also be seen is that the manual steps result in the demand being reduced for a longer period than the DSR event requires, with the reduction occurring 5 minutes before the event is required to commence and taking 3 minutes after the event has stopped to start returning to normal. While not a significant amount of time, this highlights how the different response times noted in Fig. 13.4 require the aggregator to start the process before the official event time to ensure the reduction is achieved in time. At the end of the event, the delay in restarting the chillers results from the aggregator operators having to respond to the stop request from the National Grid and only then trigger the stop signal for the hotel.

The ability to automate elements of the process becomes more important when looking at the function of an aggregator, which requires management of multiple sites during the same DSR event. During a normal DSR event the hotel scenario previously described is repeated across many different sites at the same time. Therefore, if the process was completely manual, it would require the aggregator to have enough operators to call the hotels within a short period and for the hotels to have the right people ready to manually turn off the chiller systems. Reliance on manually implemented steps to such an extent would increase costs, increase response time and potentially cause mistakes or missed reductions.

An example of aggregated loads is provided in Fig. 13.6, which shows actual versus expected flexibility that hotels contribute to the overall reduction target for the event. The expected reduction amount per hotel is determined by the aggregator based on the size of the chiller's maximum load and load history. However, the actual reduction provided by each hotel is different for each event because of variation in factors that impact the chiller operation. Factors such as weather conditions, occupancy levels and special events will influence the actual reduction delivered for any site. This variation between expected and actual reduction occurred for the hotel example used in Fig. 13.5 and is highlighted in Fig. 13.6 during 'Event 9'. The example hotel delivered more flexibility than required by providing 150 % of expected reduction (79 kWh instead of the expected 48 kWh). For this DSR event it helped contribute to the aggregated results being 48 % higher than required, even with one hotel (the first segment) not providing any reduction. In contrast, the previous event, 8, underperformed by 22 % because of two hotels under delivering (the second and sixth segments).

Without the current level of automation employed to deliver the aggregated load reductions shown in Fig. 13.6, the level of variation between expected and actual load reductions would be higher due to missed opportunities resulting from having to rely on manual steps for contacting hotels and needing someone present at each site able to

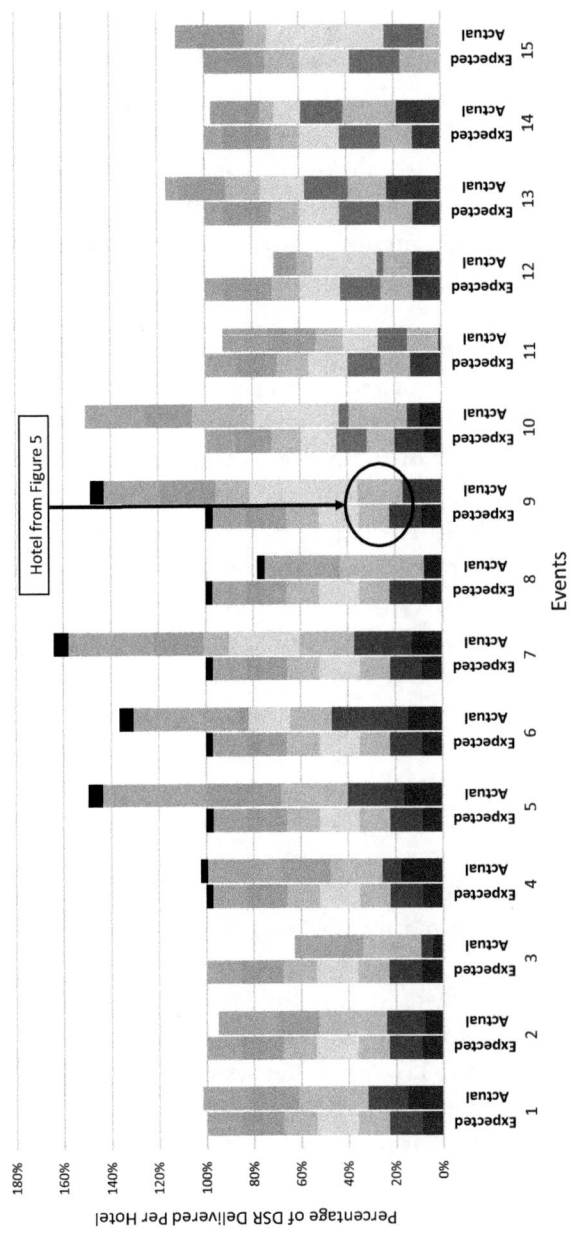

Fig. 13.6 Comparison of expected versus actual DSR load reduction for 9 hotels over 15 events

manually activate the turndown process. Therefore, aggregators prefer automated control wherever possible as this reduces the operational overhead and risk of human error. However, end users without prior experience in this process can be reluctant to let an external party control their internal systems, even for non-critical systems like chillers (Darby and McKenna 2012). This often means the aggregators need to start with manual control processes to build trust with the end user before the site's owners will authorise automated controls to be installed in the future. Responsiveness of a site is therefore not necessarily static over the life of an agreement.

The key restriction on fully automating the hotel's demand reduction process for the example used arises from the National Grid's STOR programme requirement for operators to manually confirm acceptance of the DSR event. This part of the process cannot currently be automated; however, there are other programmes that do require full automation. In the UK, the National Grid's Frequency Response programmes (FR), for example, require responsiveness in the range of sub-second to 30 seconds, depending on the service selected (National Grid 2017b). In order to meet these time frames the reduction trigger is based on changes in the electricity system frequency, not the decision of an operator at the National Grid. This rapid response requirement effectively means that all responsive loads using this programme require automated controllers that monitor the system frequency and activate when set parameters are reached (normally when the system frequency drops below 49.7 Hz). While STOR has the potential option of manual control, FR forces the need for automation. Figure 13.7 illustrates the FR interaction steps and highlights how this is primarily automated at the end user level.

The problem that the FR services present for hotels and other similar businesses is clearly the fast response time requirement, but issues of trust and control of technology use in relation to customer service are also raised. As noted in Sect. 13.6, the chiller systems have an average response time of 7.5 minutes, which is 700 % slower than is needed for participation in FR DSR services. Even if it was possible to use this service, participation would still require a new level of trust from the hotels as there is no option of being excluded from an event that is just about to start. The only way of not participating in FR is by pre-agreeing timeframes with

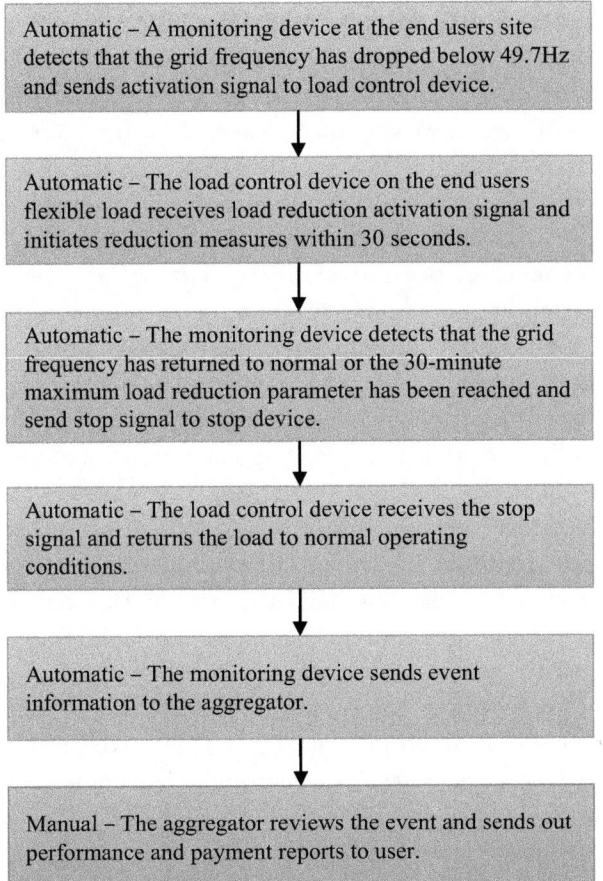

Fig. 13.7 FR demand reduction manual and automatic process flow steps

the aggregator during which the automatic starting function is disabled and instead requires additional involvement and planning from each hotel's facilities staff. So, while all loads could technically be automated for DSR, the reality is that the level of automation will be determined by the following factors: (i) trust of the loads owners to release control to a third party; (ii) technical ability of the loads to be turned down without human intervention; (iii) the flexibility of the loads to allow for reduction without impacting end users; and (iv) the responsiveness of the loads to meet DSR programme requirements.

13.7   Discussion and Conclusion

This chapter has explored issues of flexibility and responsiveness of electricity demand from an in-situ appliance and device point of view and the type of automated vis-à-vis manual intervention associated with DSR. It has applied proposed understandings of flexibility and responsiveness that seek to recognise the contextual, material and social dimensions of these characteristics of demand, in place of the rather narrow and limited understandings applied in behavioural and economic approaches.

A simple overview of different energy services in hotels showed that the air-conditioning systems in principle have the highest load flexibility and this, in turn, depends on issues of norms of user comfort, thermal inertia, controllability and size. The load characteristics of lighting, office equipment and catering and how these fit into the rhythms and demands of hotel life and the hotel as a service provider to customers are critical in understanding why these have a lower flexibility and yet higher responsiveness than HVAC. It is concluded that in hotels, a request from the system (i.e. from the Transmission System Operator first and, consequently, from an aggregator) to temporarily reduce electricity demand is implemented following a combination of technical understanding of loads and their social underpinnings. Hotels are therefore an example of how flexibility and responsiveness are strongly determined by the technical arrangements associated with demand management and the scope for intervention into sets of already interwoven and locked together practices of hotel life. This is in contrast with the dominant approach in the social sciences, which sees direct feedback and price elasticity as key determinants of flexibility and responsiveness. A more thorough understanding of how DSR occurs and which loads contribute to flexibility consists of engaging with particular loads and their technical and social features.

Through DSR there are possibilities for service providers, aggregators and end users to materialise demand shifting either with a high level of human intervention or through automation. This chapter provided a breakdown of possible areas of automation in STOR, which is the main DSR programme in the UK. Despite concerns around social acceptability, automation can be seen as having a multiplier effect on flexibility and, even more significantly, on responsiveness, because it enables rapid shifting of the timing of demand without the physical (or manual) intervention of any human actors. What is more, automation is indispensable for DSR programmes where demand frequency needs to be changed within seconds or fractions of seconds. Flexibility by technology (i.e. without changes in people's ongoing doings) can also affect the temporality and spatiality of energy demand. Can demand shifting be invisible so that little or no noticeable modification happens to demand volumes? Whilst some demand can be moved in time by way of load shifting and turndown (e.g. relaxing temperature and humidity control set points of heating ventilation and air-conditioning systems (Fasiuddin et al. 2010)), the ability to enact these changes at the required time and the ability to shift load on the system with no impact on end users depend on different technologies, the services they provide and how central they are seen to be to end users' expected or assumed needs and experiences.

The research work presented in this chapter has two main implications for practitioners and policy makers.

The first implication consists of the distinction between automation and manual intervention in DSR programmes. Different levels of investments and social acceptability are associated with these two modalities of DSR. In the US for a couple of decades temporal shifting of energy demand via DSR took place thanks to telephone communications and physical intervention by energy managers (Torriti 2015). The example of STOR for hotels in this chapter illustrates how at the moment in the UK the co-existence of automation and manual intervention is performed by practitioners. Improvements in ICT technologies and diminishing costs may lead to a higher level of automation, but this depends, among other things, on the age of devices and appliances as well as the issue of trust and loss of local control. Policy initiatives to foster automation need to be confronted with current practice and potential resistance to control delegation. For instance, a UK Government initiative to install sensors to new fridges so that these could be switched on and off remotely during times of peak demand to reduce the load on the electrical grid was revisited following negative media exposure about concerns over privacy (Spence et al. 2015).

The second implication regards detailed knowledge about which load profiles are responsive and how this may inform DSR activities. How aggregators choose customers and operate DSR may be affected by issues of size of potential demand shifting from flexible technologies (including storage and stand-by generation), but in the future may need a more thorough investigation about the types of loads and their responsiveness within the settings in which they exist. This will be particularly relevant if policies start discriminating against stand-by generation flexibility (e.g. diesel) in favour of load turndown responsiveness. DSR policy will also need to be informed by detailed contextual knowledge on which sectors should be prioritised and what expectations can be associated with DSR in a low-carbon future.

Acknowledgements This work was supported by the Engineering and Physical Sciences Research Council [grant number EP/K011723/1 and EP/G037787/1] as part of the RCUK Energy Programme and by EDF as part of the R&D ECLEER Programme.

Bibliography

Alberini, A., and M. Filippini. 2011. Response of residential electricity demand to price: The effect of measurement error. *Energy Economics* 33: 889–895.
Bandura, A. 1969. *Principles of behavior modification*. New York: Holt, Rinehart and Winston.
Barton, J., S. Huang, D. Infield, et al. 2013. The evolution of electricity demand and the role for demand side participation, in buildings and transport. *Energy Policy* 52: 85–102.
Blázquez, L., N. Boogen, and M. Filippini. 2013. Residential electricity demand in Spain: New empirical evidence using aggregate data. *Energy Economics* 36: 648–657.
Bradley, P., M. Leach, and J. Torriti. 2013. A review of the costs and benefits of demand response for electricity in the UK. *Energy Policy* 52: 312–327.
Broberg, T., R. Brännlund, A. Kazukauskas, et al. 2015. *An electricity market in transition – Demand flexibility and preference heterogeneity*. Eskilstuna: Centre for Environmental and Resource Economics, Umeå School of Business and Economics, Umeå University. Available at: http://ei.se/Documents/Publikationer/rapporter_och_pm/Rapporter%202015/Rapport_An_electricity_market_in_transition_Umea_universitet.pdf

Buryk, S., D. Mead, S. Mourato, et al. 2015. Investigating preferences for dynamic electricity tariffs: The effect of environmental and system benefit disclosure. *Energy Policy* 80: 190–195.

Butcher, K.J. 2012. *CIBSE guide F – Energy efficiency in buildings*. London: CIBSE.

Carrier. 2014. Carrier energy demand system provides automated energy management. *Carrier*. Available at: http://www.carrier.com/carrier/en/us/news/news-article/carrier_energy_demand_system_provides_automated_energy_management.aspx. Accessed 2 Feb 2016.

CIBSE. 2015. *CIBSE guide A – Environmental design 2015*, Lavenham: CIBSE

Competition and Markets Authority. 2016. *Energy market investigation: Provisional decision on remedies*. London: Competition and Markets Authority. Available at: https://assets.publishing.service.gov.uk/media/56efe79040f0b60385000016/EMI_provisional_decision_on_remedies.pdf

Darby, S.J., and E. McKenna. 2012. Social implications of residential demand response in cool temperate climates. *Energy Policy* 49: 759–769.

DECC. 2014. *Developing DECC's evidence base*. London: Department of Energy & Climate Change. Available at: https://www.gov.uk/government/uploads/system/uploads/attachment_data/file/270126/FINALDeveloping_DECCs_Evidence_Base.pdf

Dolman, M., I. Walker, A. Wright, et al. 2012. *Demand side response in the non-domestic sector*. Cambridge: De Montfort University. Available at: https://www.ofgem.gov.uk/ofgem-publications/57014/demand-side-response-non-domestic-sector.pdf

Espey, J.A., and M. Espey. 2004. Turning on the lights: A meta-analysis of residential electricity demand elasticities. *Journal of Agricultural and Applied Economics* 36: 65–81.

Fasiuddin, M., I. Budaiwi, and A. Abdou. 2010. Zero-investment HVAC system operation strategies for energy conservation and thermal comfort in commercial buildings in hot-humid climate. *International Journal of Energy Research* 34: 1–19.

Grein, A., and M. Pehnt. 2011. Load management for refrigeration systems: Potentials and barriers. *Energy Policy* 39: 5598–5608.

Grünewald, P., and J. Torriti. 2013. Demand response from the non-domestic sector: Early UK experiences and future opportunities. *Energy Policy* 61: 423–429.

Hong, J., C. Johnstone, J. Torriti, et al. 2012. Discrete demand side control performance under dynamic building simulation: A heat pump application. *Renewable Energy* 39: 85–95.

Karlin, B., R. Ford, and C. Squiers. 2014. Energy feedback technology: A review and taxonomy of products and platforms. *Energy Efficiency* 7: 377–399.

Krishnamurthy, C.K., and B. Kriström. 2013. Energy demand and income elasticity: A cross-country analysis. *CERE Working Paper* 5: 30.

Lutzenhiser, L. 1993. Social and behavioral aspects of energy use. *Annual Review of Energy and the Environment* 18: 247–289.

Macalister, T. 2015. Marriott hotels using energy demand reduction to cut carbon footprint. *The Guardian*. Available at: https://www.theguardian.com/environment/2015/may/06/marriott-hotels-using-energy-demand-reduction-to-cut-carbon-footprint. Accessed 2 Mar 2017.

Mattioli, G., and J. Anable. 2017. Gross polluters for food shopping travel: An activity-based typology. *Travel Behaviour and Society* 6: 19–31.

Mudie, S., E.A. Essah, A. Grandison, et al. 2016. Electricity use in the commercial kitchen. *International Journal of Low-Carbon Technologies* 11: 66–74.

National Grid. 2017a. *Balancing services*. Available at: http://www2.nationalgrid.com/uk/services/balancing-services/

———. 2017b. *Frequency response services*. Available at: http://www2.nationalgrid.com/uk/services/balancing-services/frequency-response/

Nicholls, L., and Y. Strengers. 2015. Peak demand and the 'family peak' period in Australia: Understanding practice (in) flexibility in households with children. *Energy Research & Social Science* 9: 116–124.

Ofgem. 2017. *Energy demand research project*. Available at: https://www.ofgem.gov.uk/gas/retail-market/metering/transition-smart-meters/energy-demand-research-project

Poudineh, R., and T. Jamasb. 2014. Distributed generation, storage, demand response and energy efficiency as alternatives to grid capacity enhancement. *Energy Policy* 67: 222–231.

Proffitt, E. 2016. *Profiting from demand side response*. Major Energy Users Council & National Grid. Available at: http://powerresponsive.com/wp-content/uploads/2016/11/ng_meuc-dsr-book.pdf

Shove, E., and H. Chappells. 2001. Ordinary consumption and extraordinary relationships: Utilities and their users. In *Ordinary consumption*, ed. J. Gronow and A. Warde, 45–58. London: Routledge.

Siano, P. 2014. Demand response and smart grids—A survey. *Renewable and Sustainable Energy Reviews* 30: 461–478.

Silva, V., V. Stanojevic, M. Aunedi, et al. 2011. Smart domestic appliances as enabling technology for demand-side integration: Modelling, value and drivers. *The Future of Electricity Demand: Customers, Citizens and Loads* 2011: 185–211.

Skinner, B.F. 1938. *The behavior of organisms: An experimental analysis*. New York: Appleton-Century-Crofts.

Spence, A., C. Demski, C. Butler, et al. 2015. Public perceptions of demand-side management and a smarter energy future. *Nature Climate Change* 5: 550–554.

Strbac, G. 2008. Demand side management: Benefits and challenges. *Energy Policy* 36: 4419–4426.

Torriti, J. 2015. *Peak energy demand and demand side response*. London: Routledge.

Torriti, J., R. Hanna, B. Anderson, et al. 2015. Peak residential electricity demand and social practices: Deriving flexibility and greenhouse gas intensities from time use and locational data. *Indoor and Built Environment* 24: 891–912.

Vallacher, R., and D. Wegner. 1987. What do people think they are doing? The presentation of self through action identification. *Psychological Review* 94: 3–15.

Walker, G. 2014. The dynamics of energy demand: Change, rhythm and synchronicity. *Energy Research & Social Science* 1: 49–55.

Xue, X., S. Wang, Y. Sun, et al. 2014. An interactive building power demand management strategy for facilitating smart grid optimization. *Applied Energy* 116: 297–310.

Mitchell Curtis is a Research Engineer in the Technologies for Sustainable Built Environments Centre, University of Reading, UK. Prior to becoming a Research Engineer, Mitchell spent over 10 years working in the information technology and communications industries as an analyst and project manager. Mitchell has a bachelor in commerce from Canterbury University, a master of communications from Victoria University, a master of science in Renewable Energy from University of Reading, and is currently undertaking a PhD in engineering.

Jacopo Torriti is a Professor in Energy Economics and Policy in the School of the Built Environment, University of Reading, UK. Before joining the University of Reading in 2011, Jacopo held teaching and research positions at the London School of Economics, the University of Surrey, the European University Institute and the Massachusetts Institute of Technology. He obtained a PhD from King's College London, a master in European Studies from King's College London and a laurea in economics from Università di Milano.

Stefan Smith is a Lecturer in Energy Systems and the Built Environment in the School of the Built Environment, University of Reading, UK. Stefan took up this position in 2012, having been a research fellow at the Institute for Energy and Sustainable Development, De Montfort University, researching energy system modelling and implications of climate on energy system behaviours. He has a PhD from the University of Nottingham, a masters in software and systems development from the University of Glasgow, and a BSc (Hons) in physics from the University of Nottingham.

14

Reducing Demand for Energy in Hospitals: Opportunities for and Limits to Temporal Coordination

Stanley Blue

14.1 Introduction

This chapter argues that demand for energy in institutions is the outcome of changes in various forms of institutional organisation of working arrangements. Intervening in the organisation and timing of working arrangements is presented as an opportunity for both improving demand response as well as reconstituting working arrangements in such a way that they demand less energy. I argue that opportunities to steer demand in this way, through temporal coordination, depend on there being sufficient flexibility in the various forms of connection between working practices that hold temporal arrangements in place.

To make this argument, I refer to three empirical examples taken from a study of three acute[1] National Health Service (NHS) hospitals. I refer to hospitals as example cases for two reasons. First, the NHS is the largest public sector contributor to climate change in Europe and one of the largest employers in the world. Through the running of its services it is

S. Blue (✉)
Lancaster University, Lancaster, UK

© The Author(s) 2018
A. Hui et al. (eds.), *Demanding Energy*, DOI 10.1007/978-3-319-61991-0_14

responsible for approximately 25 million tonnes of carbon emissions each year and for the organisation of the working lives of 1.7 million people (see NHS Sustainable Development Unit 2016). Predictions estimate that even if the NHS successfully delivers all of its currently planned interventions, it will still miss its government set 2050 carbon emissions reduction target by 26 %. The NHS represents an example of a large and complex organisation that requires significantly new ways of thinking about and reducing its energy consumption. The second reason for referring to hospitals as example institutions is that the hospital has already been well studied as an institution that makes and shapes patterns of working. In this chapter, I draw on and extend Zerubavel's (1979) ethnographic study of hospital life in the US. The example cases that I refer to are taken from interviews that were part of my own ethnographic study of institutional rhythms and the organisation of working practices in hospitals. This research was conducted over a six-month period at three sites and included observations of daily working practices and routines in different departments, from auxiliary and background services such as laundry and decontamination departments to frontline clinical services delivered on wards, in operating theatres and in intensive care units. Twenty-seven interviews were held with a range of staff from hospital and estates managers, to nurses and clinicians, and to facilities and operational staff. I use this material to illustrate a way of thinking about how demand is made in institutions. I do not claim that this way of thinking can be inferred by the narratives presented. Rather the example cases help me to show up and describe how healthcare services have changed, are changing and do not change, and how these changes and obduracies matter for the constitution of demand for energy in hospitals.

From here I go on to say more about the current NHS strategy for reducing carbon emissions to show how an understanding of how demand for energy is made is missing. In the following section, I describe how working arrangements in hospitals are organised temporally, and in other ways. I then go on to illustrate, through an example of the changing role of diagnostics in the provision of health care, how the timing of working practices depends on interconnections between temporal and material arrangements. In the final two sections, I show how potential changes to the timing of working arrangements, as a strategy for managing and reducing demand for energy, depend on there being sufficient flexibility across various forms of connection between working practices. I illustrate this with examples referring to the rearrangement of the way that breast cancer services are delivered in the NHS and the limitations on clinical developments in radiology services.

14.2   What Do Hospitals Use Energy for?

Reductive accounts of consumption, and of energy use, underpin current NHS sustainable development strategies, limiting potential opportunities for managing and reducing demand for energy.[2] Demand for resources like energy, as well as the services that those resources are for, are considered to be made outside of the hospital, to be disconnected from everyday hospital life, and to be something that the hospital has little control over and has to respond to accordingly. The idea that demand is made externally and met through the delivery of services is reflected in the NHS Sustainable Development Unit's strategy for reducing carbon emissions (see NHS Sustainable Development Unit 2016). On the one hand, the strategy counts on, and expects, the health sector and hospitals to receive carbon savings as a result of changes in national and international government policy, including changes in: "… public spending, increased renewable energy in the UK energy mix, improved carbon efficiency of production and more fuel efficient vehicles." (NHS Sustainable Development Unit 2016: 7). On the other hand, the actions it proposes that the health sector itself should take to meet demand for resources in a less carbon-intensive way include: adopting more efficient technologies, procuring products made in less carbon-intensive ways, and implementing energy and waste saving programmes that require changing the behaviour of staff. I leave criticisms of this kind of approach to energy 'efficiency' (see e.g. Wallenborn 2015), of the promise of renewable energy sources, and of a dependence on technological innovation for carbon reduction to one side in this paper, and instead highlight that there is no discussion of how the health sector itself is implicated in the making of demand for energy. Energy consumption is instead taken to be steadily increasing and the only question, for those interested in sustainable development within the NHS, is one of how to supply this growth in the least carbon-intensive way. However, consumption is not static, it is not permanently and unilaterally growing, and there are more questions to ask and to answer beyond how to keep up with growing demand. Demand is rather dynamic, changing in different ways all the time. There are a wealth of social scientific ideas that can help to study and understand these dynamics and that can help us to understand possible new opportunities for steering and shaping demand.

Within the social sciences, a strong case has been made for treating patterns of consumption (Warde 2005), and particularly of energy consumption (Shove and Walker 2010), as outcomes of practices and the ways that practices are arranged. In their article, 'What is energy for?', Shove and Walker (2014) challenge the idea that energy supply and demand is either a cause or a consequence of changing political, economic and technological systems. Instead, they propose an approach that views energy demand as constituted by practices, or by what people do. In their words:

> To persistently ask 'but what is energy for?', and to take that as the central question is to take a different view of the social. It is to see society not as an outcome of intersecting systems, like geological forces pressing this way and that, but as emergent from, and defined by, social practice. (Shove and Walker 2014: 6)

Rather than conceiving of energy use as the result of tectonic shifts in national and international policy, in changing economic strategies, and of technological and scientific progress, these authors call our attention to the practices that energy is for. Instead of focussing on limiting and enabling social structures that govern the consumption of energy, an approach that focuses on practices draws our attention to the social reproduction of everyday activities at home and at work that constitutes what are seen as normal and acceptable ways of life and that are what energy is used for. In the hospital, practices might include the prescribing, distributing and taking of medicine; diagnosing, testing and treating patients; as well as eating, washing and resting. While different theoretical traditions reflect different definitions of a practice (see Schatzki 2016), what practice theorists have in common is that they argue for the centrality of practices in the analysis of the constitution of social life and how it changes.

Energy demanding practices are not free to be reproduced at any time and in any place. Instead, they are connected in time and across space, and it is this spatiotemporal extension that facilitates their effective and regular reproduction. They can be connected in sequence. For example, diagnosing, prescribing, ordering, receiving and then, administering medicine. They might also be connected in space. For example in hospitals there are a range of activities taking place at any one time: people being treated, having operations, taking bed rest; people serving and drinking coffee in the café; managers having strategic, planning and investment meetings; staff delivering goods, cleaning up, serving lunch, and so on. The idea that practices connect and that they are interdependent is important for understanding how demand is made. The reproduction of a given practice can require the reproduction or indeed exclusion of another practice, or set of practices. Cancelling a clinic, for example, might mean that a whole range of connected activity including patient travel, follow-up appointments, tests and admissions do not happen. While admitting a patient requires a whole host of administrative, diagnostic and treatment practices to be performed. The central challenge of reducing energy demand, for authors such as Shove (2009), Shove et al. (2012), Southerton et al. (2011), and Walker (2014), is one of modifying practices and their arrangements so as to disrupt the regular reproduction of practices and reduce consumption associated with practices that might not be enacted.

In this chapter, I work with these ideas to challenge prevailing understandings of how demand for energy is made in institutions like hospitals and more broadly in large organisations like the NHS, by focussing on the practices that hospitals use energy for. Because hospitals use energy in the regular provisioning of healthcare services, and to explain how hospitals make demand, I go on to describe how working practices involved in the delivery of healthcare are organised, shaped and changed by various forms of emerging connections between practices.

14.3   How Hospital Life Is Organised (Temporally and in Other Ways)

One way that working practices are connected, and hospital life (and hence energy demand) is organised, is temporal. Although hospitals are often perceived to be highly irregular organisations that respond to emergencies and to changes in patients' health, as well as to changing demographics and diseases, in many ways hospital work is highly ordered and routinised. This is because it is underpinned by a socio-temporal structure that shapes 'normal' and acceptable ways of working. In his ethnographic study *Patterns of Time in Hospital Life* (1979), Zerubavel describes the temporal sequences and socio-temporal cycles that run through and structure the organisation of hospital working arrangements. He argues that sequences of working practices in departments and across the hospital make up socio-temporal cycles that structure the organisation of hospital life. These include the duty period, shift patterns and staff rotations which intersect with social cycles of the day, the week and the year, as well as with cycles of career progression and training. The culmination of intersecting sequences and cycles is that working practices in hospitals exhibit a rhythmic temporal pattern. In Zerubavel's words:

> [Both]… routine and nonroutine events and activities [are forced] into regular temporal patterns, thus introducing a rhythmic structure into hospital life… even purely medical events and activities are forced into rhythmic patterns which are dictated by nonmedically based schedules. (Zerubavel 1979: 34–35)

The timing of a given practice is, therefore, a product of its connections to other practices, as well as of the temporal organisation of the hospital as a whole. This holds true of both medical and non-medical activities. Zerubavel explains, for example, how it is purely a social convention that medicines are distributed and therefore required to be taken every 4 hours rather than every 4 hours and 18 minutes. This timing is a product of the close sequencing of medicine rounds with meal times (1979: 35).

Connections between practices and their interdependencies mean that certain practices occupy central and fixed positions in practice arrangements and therefore in hospital schedules, while others are more flexible and can happen at different times. Meal times on wards, for example, act as a strict temporal anchor around which other practices are arranged, while completing paperwork, writing up patient notes and performing clinical observations are practices which have more flexible positions in daily schedules. Southerton describes how the temporal organisation of daily life matters for the fixity and flexibility of different practices. He writes that:

> The temporal organisation of the day can be characterised as being constituted by practices that have a fixed position within schedules... These are surrounded by interrelated practices that have a more malleable position within sequences, leaving a stock of practices contingent on filling empty slots within the day. (Southerton 2006: 451)

Zerubavel and Southerton help us to see that working practices are connected temporally and that the temporal organisation of hospital life matters for the fixity and flexibility of practices in hospital schedules. These are important observations for understanding how institutions make patterns, profiles and peaks in demand for energy. If demand is the outcome of the regular reproduction of working practices and practices are organised temporally, then both managing and reducing demand depend on reconfiguring the temporal organisation of institutions and on the making and breaking of connections between practices.

On the one hand shifting the timing of working arrangements is a way of achieving demand side response. Demand side response has gained increased attention from system operators and regulators because of its potential to (1) lessen the effects of high demand on ageing infrastructure; (2) reduce requirements to draw off more carbon-intensive supplies and hence reduce costs to consumers; (3) and to better match the timing of demand with the timing of outputs from low carbon generation (wind, solar, etc.). Peaks in energy consumption are also likely to become more problematic in the future as services like heating and mobility move to electrification (Powells et al. 2014). Changing the timing of energy-intensive activities so that they take place outside of peak demand is one strategy for reducing energy consumption that requires a careful understanding of the timing of practices and their potential fixity and flexibility in hospital schedules.

However, this uneven distribution of demand is also representative of the temporal patterning of total demand that underpins the range of practices that make up hospital life. While peaks are the outcomes of societal synchronisations, the outcomes of many energy-intensive practices happening at the same times, societal synchronisations are held in place by the same socio-temporal orders and rhythms that Zerubavel describes. Southerton, writes: "Peak loads in energy consumption and transport represent simple empirical observations that reveal such rhythms." (2013: 344) The temporal order that is revealed by observations of these kinds of societal synchronisations is multi-scalar, produced by the activities that are performed daily, weekly and annually. Walker argues that: "Each of these scales of rhythms in practice – daily, weekly, seasonal – and their interaction, are generative of the rhythmic patterns of energy demand." (2014: 51) The rhythmic patterns of energy demand are underpinned by the socio-temporal ordering of practices. Torriti follows this line of thinking when he argues that the significance of peak load is that it is representative of the way that practices are ordered in time:[3]

> Peak energy demand emerges as a phenomena which epitomises the relevance of practices as a unit of analysis in this context. At the heart of the approach which places social practices at the centre of our understanding of the dynamics of energy demand is the position that the timing of energy demand is determined by the way practices are ordered in time. (Torriti 2017: 37)

Shifting the timing of working arrangements, therefore, is not only a strategy for desynchronising energy-intensive practices so that peaks are shaved, but it also reconfigures the temporal organisation of working practices, and hence total load. As a result, energy load is not just displaced through the changing timing of working practices; it can also be dissipated as temporal connections are broken, and previously connected practices are no longer reproduced. Following Zerubavel, problematic and unevenly distributed profiles of demand, as well as patterns that represent total demand, can, therefore, be understood as the outcomes of problematic or conflicting social cycles. Opportunities for resolving these can be found in strategies of temporal coordination, in intervening in and shaping the temporal organisation of the hospital. But opportunities for modifying the temporal organisation of working practices to shift peaks and modify patterns of demand so that they are less resource intensive depends on the flexibility of specific temporal connections and the way that those arrangements are held in place by other forms of connection between working practices.

Hospital life is not only organised temporally. Practices also connect, and arrangements are underpinned, by other forms of connection that matter for the temporal organisation of hospital life. In previous work (Blue and Spurling 2016), we suggested that the temporal organisation of the hospital *inter*connects with professional boundaries and material arrangements,[4] as well as with other forms of connection between practices. Professional boundaries, or as we have described them, jurisdictional connections (the abstract organisation of expertise and division of labour), that matter for who works when, where and with whom, are reflected in the temporal sequences and cycles in the hospital. Similarly, material connections between practices interconnect with temporal and jurisdictional ones. The physical location of certain departments, for example, embedded in hospital design, permits and limits opportunities for collaboration, working in sequence, sharing equipment, and so on and holds in place (literally) particular ways of working.

Recognising that working practices in the hospital are not only arranged temporally but also by other forms of connection matters for understanding the fixity and flexibility of working practices and hence for opportunities for temporal coordination to disrupt and reconfigure patterns of consumption, to shift and reduce demand. Instead of seeing fixity or flexibility as a given characteristic of a practice, or a product of its positioning within the intersecting temporal sequences and social cycles, the potential for shifting the timing of working practices rather depends on the historical layering (Blue and Spurling 2016) of connections between practices and interconnections between various forms.

In describing how past practices and their arrangements matter for present and possible future configurations, Schatzki writes:

> …the pasts of practices, arrangements, interwoven timespaces, and social phenomena are among the items that circumscribe, induce, and underwrite the public presence of activity – and, thus, their own emergence, persistence, and transformation. The practices, arrangements etc. that arise from human activity in turn contextualize activity. History thus embraces mutually dependent activities, arrangements, practices and social phenomena. (Schatzki 2010: 214–215)

Although Schatzki is working with a specifically developed schema of practice theory here, the argument is that past arrangements of practices, in their temporal, material, jurisdictional and other forms, circumscribe, induce and underwrite the emergence, persistence and transformation of practice arrangements. It follows, therefore, that fixity and flexibility of practices are a product of those emerging forms of connection. Opportunities for temporal coordination are dependent on the historical layering and emergence of interconnected forms of material, temporal and jurisdictional connections.

In the following three sections, I illustrate how the historical layering of interconnections matters for emerging fixities and flexibilities through a description of three examples: emerging interdisciplinary working in breast cancer services in the NHS; fixed temporal working arrangements in radiography; and changing material arrangements in pathology.

14.4   Changing Material Arrangements in Pathology

Innovations in technology are not straightforwardly responsible for increasing demand for energy. Neither does innovation in healthcare technology drive the reconfiguration of working practices in the hospital. In some cases configurations of working arrangements can be so strongly embedded that technologies will be rejected should they significantly disrupt the temporal or jurisdictional order. It is instead the historical layering of interconnections which afford flexibilities in types of connections and establishes new limits on possible future ways of working. The role of pathology services in the hospital (referred to sometimes by hospital staff using the umbrella term diagnostics which includes pathology, radiology, and other kinds of testing services) has changed dramatically over the last 40 years, in part because of changing technological capabilities. Lawrence, a manager for pathology services, described this change and some of the reasons for it:

> Diagnostics thirty to forty years ago was used very much as a confirmation of a diagnosis which they'd [doctors] already arrived at. So they did the physical examination of the patient, took the history of that [patient] and said I think this is what's wrong with you, we'll do the test to confirm. So that's the philosophy. Unfortunately, now it's reversed on its head, diagnostics is being done upfront. So the diagnostic is driving the decision-making process. (Lawrence, pathology services manager)[5]

While pathology had previously provided confirmation of diagnoses that consultants and doctors arrived at through other kinds of medical tests, energy-intensive services like pathology and radiology now form part of the services provided at the very beginning of many patient pathways.[6] Instead of featuring more sparingly as a confirmation tool, sometimes after treatment had already begun, or even when a patient had left the hospital, diagnostic services are now used to identify specific pathologies and set in motion courses of action to treat the patient. So while resource-intensive activity in the lab and in radiology had previously been part of some, but not all, patient pathways, it now forms a part of the beginning of a majority of sequences of activity in the hospital.

Lawrence provided various explanations for this reversal, for the changing role of diagnostics, for the expansion of its services and its increasing forms of dependence and connections to a wide range of clinical services. One explanation was that services in pathology had to respond to increasing demand made externally by shifting economic, political and technological landscapes. He suggested that pathology services had developed in line with medical and technological advances. As medicine has been able to specify in more detail exact cancer sub-types, diagnostic services have had to keep up. In Lawrence's words:

> [I]n certain areas the work and the complexity of the work that we're doing is very much being driven by... for example in histopathology... we are needing to type the tumour in more detail... because of designer drugs, designer cancer drugs. [C]ertain cancer drugs are [now] being designed [to a] very... specific tumour subtype. In order to subtype that tumour you need to do a significant amount of work. So a lot of that sort of diagnostic work is now being done upfront... (Lawrence, pathology services manager)

Lawrence partly puts this shift in demand for testing down to external advances in medical research and technology, following a prevailing narrative and one that is followed by various organisations including the NHS Sustainable Development Unit. However, he went on to describe other aspects to this growth in demand for pathology services, which were about connecting pathology with different services across the hospital. As pathology services became more developed, they have forged dependencies with different clinical services, and become intimately integrated with new forms of treatment and care.

> The other element of it is very much around therapy and management of the patient. A lot of the new therapeutic agents, in particular, chemotherapy... require certain tests to be done... to demonstrate that the patient is responding to that treatment, but some of those drugs are quite toxic in their own right [so] it is to also monitor the side effects. To make sure that [the side effects are] within acceptable limits and... not... causing an issue for the patient. So there is that element of it. (Lawrence, pathology services manager)

Lawrence complicates the narrative that medical and technological advances directly relate to demand for pathology services, and hence to the demand for energy and other kinds of resources. The delivery of new treatments and services is partly enabled by changes in the roles, remit and capacities of diagnostic services. The development of new chemotherapy treatments does not develop independently from changes in diagnostics, but emerges symbiotically. As Lawrence himself reflects: "[It] is… an extremely interrelated, multi-factorial impact in terms of what's actually… driving the demand [for pathology services]."

The changing role and remit of pathology services have of course been part of improving patient treatments and overall population health. It has also impacted demand for energy across the healthcare service. The availability of a wider range and more detailed forms of testing has increased the number of tests being performed and increased electricity consumption as well as the fuel required to transport samples (often individually) for testing. Pathology services are usually at their busiest in the late afternoon and evening peak period. This is because those services receive the bulk of samples in the late afternoon following the accumulation of sample and specimens that have been drawn from clinics during the day and as a result of newly implemented turnaround times. In this way, material arrangements connect practices in ways that matter for and at the same time have to fit into the temporal organisation of the hospital.

Advances in the technological capabilities of diagnostic services modify the sequence of practices of diagnosing, treating and testing. But this modification also has significant implications for the temporal configuration of practices and the potential organisation of working arrangements. Changes in this sequence has resulted in increased testing and a fixing of times when pathology services operate (in sequence and coordinated with a wide range of clinical activities). And at the same time this modification in sequence, facilitated by new material forms of connection, has afforded pathology services the opportunities to connect with, integrate with, and underpin all kinds of other services. The ability to test at faster rates, more accurately, and for a broader range of diseases has enabled a culture of testing and, for some, a dependency on pathology services. It has provided the ground for the emergence of new clinical practices that themselves require more testing, and embedded testing in existing clinical procedures. Perpetuating this demand for testing and dependence on services in pathology has immediate energy consequences, but it also has broader implications for length of stay and 'bed flow' while patients wait for test results, as well as for the increased demand for energy and other resources that accompany patient stay.

The example of changing material arrangements in pathology services illustrates that demand for a range of healthcare services and related energy is not straightforwardly an outcome of the adoption of and integration of new technology, but rather the outcome of changing interconnections between material and temporal connections between practices. New material arrangements allow faster provision of testing and allow pathology to be connected in different ways to clinical practices and to become embedded in new kinds of practices. There are also other forms of connection beyond the material and the temporal that matter for the organisation of hospital life and hence for how the hospital makes demand for energy.

14.5   Flexible Professional Boundaries in Breast Cancer Services

Material arrangements are not only interconnected to the temporal organisation of the hospital, but also to professional boundaries that shape ways of working. The capacity for temporal coordination then is tied to the flexibility of material, jurisdictional and other forms of connection. One example that shows well the importance of this historical layering of interconnections between practices that circumscribes, induces and matters for the fixity and flexibility of practices in arrangements, and hence temporal coordination and peaks and patterns of energy demand, is the rearrangement of the way that breast cancer services are delivered in the NHS.

NHS hospitals now run what is known as a one-stop breast clinic for patients referred by their GP for breast cancer symptoms. Before these clinics were established, patients with such symptoms were required to have a series of appointments in different departments across the hospital (and sometimes in different hospitals), with potentially long waiting times in between, before they could get a diagnosis. Now patients receive all of this care in a one-stop clinic, on the same day, and in the same session. Requirements were introduced by the Department of Health in 2010 that all breast patients are required to be seen by a consultant with all accompanying test results within two weeks of referral. This policy and set of temporal targets are stringently adhered to by NHS hospitals and is enforced through the use of financial penalties when targets are 'breached'. This policy, however, did not impose a fundamental rearrangement of hospital services, instead working practices had been developing in this way for some time, and government policy rather codified and standardised already emerging ways of delivering breast cancer services.

In an interview with Elliot, a surgeon and clinical lead for breast services, we discussed the historical developments and transformations that led to closer and new kinds of working arrangements between the services involved in diagnosing and treating breast cancer. This discussion shows up the historical layering of material and jurisdictional connections between practices that matter for the temporal organisation of these services and the possibility of the two-week target. Elliot described some of the changes that allowed breast cancer services to be delivered in the fashion of a one-stop clinic for patients. She first explains that the integration of new technology reduced the duration of time that it took to test for cancers and that this allowed services to come together on a single day. In Elliot's words:

Elliot	[I]t was the introduction of fine needle cytology. So you had the opportunity to be able to provide a pathology result on the day of the clinic appointment. So the patient could have the, we call it a triple examination that people get in breast clinic, a clinical examination, the radiology, and the pathology. And when these one-stop breast clinics first started people were getting the examination, the imaging and… the cytology [in one day].
Interviewer	So the availability of that new technology is what allowed those services to be able to come together?
Elliot	Yes and that's going back about 30 years I think… people were doing cytology then and setting up one-stop breast clinics. And they just became the norm; breast clinics were just done that way. (Elliot, breast cancer services manager)

This new and 'normalised' arrangement represents a shift in the temporal organisation of working practices. Patients no longer have to come to the hospital three times, but only once to receive the same diagnosis that they would have 30 years ago. The departments involved no longer run independently, but have to provide working arrangements and patient pathways that can cater for the one-stop clinic. As a result, timings for diagnostic test turnarounds, patient waiting times and, bed flow, are reconfigured. The frequency of patient visits, the effect on patient flow and the increased dependence on more sophisticated diagnostic methods, all impact the demand for hospital resources as well as the timings of when energy-intensive activities occur.

In bringing these services together, the one-stop clinic has advanced the ways in which breast cancer is diagnosed and treated. Closer working between radiographers, pathologists and clinicians has enabled interdisciplinary approaches, research and treatment and has led to the emergence of new kinds of tests and treatments. Elliot described the impact of bringing these services together:

> What's then happened is that the way we diagnose breast cancer has evolved from just having a cytology result into having actually a bit of tissue to report on… That then obviously takes a few days to process because it has to be fixed and blocked and looked at by the pathologist. So that's where you get into the situation where people are coming back in a weeks' time… Because [with] the cytology all you get is a cell and you can tell whether it's benign or malignant, but you can't do any other tests on it. When you get a little slither of tissue, the pathologist can then tell you what type of breast cancer it is, what grade it is, so that's how aggressive it is. And they can start to do the tests that tell you how it's going to respond to treatment as well. And all of those things have gradually become the norm, and the benchmark has moved to treating breast cancer. (Elliot, breast cancer services manager)

The close and integrated ways of working in breast cancer services have shaped the development of both the disciplines involved and the ways that they work together, and hence the opportunities and requirements for new material and temporal configurations between the working practices involved. It is only through this kind of close interdisciplinary working that requirements for more detailed sub-typing and grading of cancers have emerged and in this case, this is what has resulted in further pressure on those services, now working to a two-week target from referral to consultation.

Jurisdictional forms of connections, like material and temporal connections, can also spread across practice organisations, where the temporal and material connections that underpin them are flexible enough to accommodate this. Elliot notes that the way that breast cancer services have been rearranged have become a model for bringing together other services in hospitals to meet similar temporal and financial targets:

> We all work much more closely in teams, and that's what's evolved through this close working I think, the multidisciplinary team working on cancer. And I think it's probably fair to say that what we've done in breast surgery or what people did in breast surgery has then spread out into other specialities and the multidisciplinary team working has become standard in all branches of cancer. (Elliot, breast cancer services manager)

Interdisciplinary ways of working that bring together specialists, knowledges and equipment from different fields have become the standard for hospital work around cancer more broadly, and appear to be being taken up in other specialities as well. Emerging interdisciplinary ways of working have implications for demand for services. In this case, bringing pathologists, radiologists and clinicians together has shifted the temporal sequence and timings of working arrangements from separated services and long waiting times to a one-stop clinic, and finally to a one-week wait for cancer sub-typing. It is clear that this changing timing matters for the number of clinics, for the number of hospital visits, and for the related resources and patient transport required. The spread of interdisciplinary working could, therefore, lead to greater temporal coordination, fewer patient visits and a reduction in resources consumed during hospital stays, waits and visits.

It is clear from this example that changes in one form of connection (jurisdictional) matter for changes in others (material, temporal) and hence the organisation of working practices in the hospital. The question of which changes in one form of connection will shape others, or be accommodated by or able to shape the complex of practices that makes up hospital life is an empirical question that depends on uncovering the historical layering of connections and the fixities and flexibilities that layering and interconnection affords. It is not always the case, as so far has been described, that technological innovation and potential new material connections will be integrated into current ways of working. Particular temporal connections, for example, can be so tightly interwoven that working practices cannot be shifted to accommodate new clinical processes.

14.6   Fixed Temporal Arrangements in Radiology

Temporal connections between practices can be so embedded, so dense, that practices and the temporal arrangements that they make can resist changes in material or jurisdictional forms of connection. The examples described until now have demonstrated how the historical layering of forms of connections between practices and their interconnections matter for the ways that hospital life is arranged. Changes in material arrangements bring practices together, shifting professional boundaries and therefore the potential for new connections between practices. For example, it was fine-needle cytology which helped to establish interdisciplinary ways of working in breast cancer services. Material forms of connection also shape temporal configurations of working arrangements. It was new material arrangements in pathology that enabled new connections and new practices to emerge.

However, the following example from radiology shows that the potential flexibility of connections between practices and hence ways of working is first a product of the historical layering of connections and second that it is relative. Complexes of practices or different forms of connection within a given complex are never in a state of fixity of flexibility, but their fixity and flexibility are relative to the kind of change in one form of connection and its impact on other types of connection. Anna, a service manager in radiology, described the various ways that the temporal organisation of working arrangements in radiology was able to accommodate certain technological changes. For example, Anna described how increased magnification of CT scanners had resulted in multiple periodic appointments and follow-ups for patients.

> ...CT scanners have gone from providing a slice through you from 10mm to sub 5 mm. What that means is the spatial resolution is increased, and we can see things that we could never see previously. So what we are finding is a lot of nodules in chests, so some people would be walking around with nodules that they don't know they've got... So we'll bring them back at six months, and we'll rescan them... And they're followed up for two years... So the equipment and the ability to see things has changed how we then follow up patients. So clearly we've found a whole load of patients that we never used to see and massively impacted on our demand yet again. (Anna, radiology services manager)

In this instance, some of the patients who were already being scanned, as a result of this increased resolution, were now being asked to return for follow-up scans at six months and two years from the original appointment. In this case, increased capacity in resolution has increased the frequency of appointments impacting on demand for services, the total number of scans and the energy required to deliver this service.

Yet Anna described that it is not always the case that working arrangements in the department are modified to incorporate additional, and potentially, lifesaving services. While the frequency of repeating this particular scan could be accommodated by extending hours or running additional clinics at the weekend, not all new services made possible through new technologies were able to be incorporated into existing ways of working. In particular, incorporating services that involved new technology and that required significant temporal reconfiguration often met with most resistance, not least because temporal boundaries were held in place by increased costs associated with weekend working and unsociable hours.

> ...cryoablations are targeted killing of liver tumours, renal tumours, and we do it by laying a patient on a CT scanner, finding the area we're interested in and then sticking needles in and ablating it, either using radio

waves, microwaves, [or] cryo freezing ... But it takes a long time, and for every hour that that patient is on a CT scanner, we could have done four outpatients. And then you start to think well if that surgeon or radiologist who does that procedure is only doing it on a Tuesday afternoon we're going to have to displace that work that would have been there and put it on a weekend. You then run into the concerns that well actually that's going to cost us more money to deliver that because at the moment weekend working is, has some enhancements on it. So there does come a point where we say we're not going to do any more of those because actually it's too expensive. (Anna, radiology services manager)

The flexibility of temporal arrangements and the possibility of temporal coordination depend on the historical layering of interconnections. In this case, temporal arrangements in radiology appear to be flexible in certain areas but not in others. It is possible to add in additional work of the same kind, but reconfiguring entire schedules is impossible, not least because of the financial limitations that accompany broader socio-temporal rhythms of the working day and week, but also because these schedules are held in place by jurisdictional boundaries. The shift patterns, working hours and rotations of surgeons, radiologists and teams of staff in radiology are not sufficiently arranged to be able to deliver this service. Reconfiguring the timing of radiology services to incorporate this new kind of treatment would require too great a shift in established ways of working.

What this example shows up is that temporal coordination as a strategy for shifting the timing of working arrangements and as a way of managing and reducing energy demand depends on the fixities and flexibilities of certain connections between practices which are afforded by a historical layering of interconnections. Understanding that changing the timing of working arrangements is limited by, for example, professional boundaries, or even further that changing material and jurisdictional forms of connection between practices has the potential to reconfigure temporal arrangements, presents new opportunities for steering and reducing demand for energy.

14.7 Conclusion

I have tried to show some of the ways that demand for energy is made in large institutions, like hospitals. I argued that hospitals do not only respond to externally made demand for services and energy but that demand is an outcome of working practices which are organised in time. I claimed that it is this temporal organisation that underpins patterns, peaks and profiles of demand and therefore, that steering demand depends on intervening in or shifting the timing of working practices to achieve what Zerubavel calls temporal coordination. However, I argued that working practices in hospitals are not only organised temporally but also through material and jurisdictional forms of connection, and that these forms of connection interconnect in shaping and holding together practice arrangements and potential opportunities for future connections. Achieving temporal coordination as a way of steering demand, therefore, depends on the relative fixities and flexibilities of interconnections.

The implication of this way of thinking about both demand side response and total demand reduction is that institutions should turn their attention from external shifts in political, economic and technological regimes, towards the ways that their own working practices are arranged. Of course, this is not to discount these kinds of systemic changes, but to understand how they are manifest in the organisation of the delivery of healthcare services. Rather than being concerned with (and depending on) advancements in renewable supply, electric cars, less carbon-intensive forms of production and changes in public spending, organisations with a remit for reducing carbon emissions in institutions like the NHS would do well to turn their attention to the healthcare services that energy is for, and to understanding their histories and the ways that they are developing. We need to know more about how different services are changing and shaping each other, and more about where they are in decline or static and contributing to or holding in place problematic profiles. Strategies for intervention, to close the projected 26 % gap in carbon savings required to meet government set targets for the NHS by 2050 that follow from this approach might include: promoting interdisciplinary working and developing services to achieve greater integration, temporal coordination and increased sharing of resources. They might include a re-skilling and reconstruction of doctors' diagnoses in such a way that they do not depend on testing in pathology, and it might include carbon measurements of potential and emerging services. Equally, such organisations would be interested in achieving more even distributions of energy (for the reasons set out above). At a minimum this would involve a systematic review of the potential for shifting auxiliary and clinical services out of problematic peak times, and of what other forms of connection (material, jurisdictional, financial) need to be rearranged to achieve this kind of temporal reconfiguration. At the most, achieving a more even distribution of demand would be tied to the project of reassessing and redesigning the organisation of a less resource-intensive healthcare service.

On the one hand, it seems counterintuitive to suggest that medical services and healthcare provision should be designed and debated with energy demand, carbon emissions and sustainability at the forefront of the agenda. Patient care is and must always be the priority for any healthcare institution. On the other hand, in this chapter, I have tried to challenge the view that medical and technological advancement happens in some way beyond the hospital and that it is more straightforwardly shaped by processes and practices in the hospital. I have shown that the provision of healthcare services is rather shaped by the flexibilities and fixities afforded by the historical layering of *inter*connections in the arrangements of working practices that makes up hospital life. Moreover, given the considerable contribution that NHS services make to total UK carbon emissions, and the rising numbers of UK hospital admissions that are attributed to air pollution and increasing extreme temperatures resulting from climate change, it is more pertinent than ever to understand more about how institutions like hospitals make demand for energy and other carbon-intensive resources and to consider new opportunities for steering that demand, perhaps through the temporal coordination of services and the reconfiguration of practice arrangements.

The suggestion to focus on how institutions shape working arrangements is not only to be applied to hospitals but could well be developed with a broader range of institutions and large organisations in mind.

Each will have its own history, its own forms of connections and its own possibilities for reconfigurations. Research could further be developed in schools, universities, prisons and other kinds of large organisations to identify significant forms of connection between practices that matter for the temporal organisation and hence energy demand of the institution, and to identify other historically made opportunities for reconfiguring patterns of work and hence for steering energy demand.

Acknowledgements This work was supported by the Engineering and Physical Sciences Research Council [grant number EP/K011723/1] as part of the RCUK Energy Programme and by EDF as part of the R&D ECLEER Programme.

Notes

1. Acute care is a level of health care in which a patient is treated for a brief but severe episode of illness, for conditions that are the result of disease or trauma, and during recovery from surgery.
2. According to the NHS Sustainable Development Unit (2013) Carbon Footprint Update for NHS in England, buildings' energy use makes up approximately 17 % of NHS carbon emissions, while transport makes up another 13 % and procurement another 61 %.
3. Peaks can, of course, be problematic in their own right, as I write above.
4. In that article we referred to these as jurisdictional and material-spatial connections.
5. Pseudonyms and generalised job titles are used throughout to preserve the anonymity of participants.
6. A typical or planned journey through the healthcare system, from first contact to completion of treatment.

Bibliography

Blue, S., and N. Spurling. 2016. Qualities of connective tissue in hospital life: How complexes of practices change over time. In *The nexus of practices: Connections, constellations, practitioners*, ed. A. Hui, E. Shove, and T. Schatzki, 24–37. London: Routledge.

NHS Sustainable Development Unit. 2016. *Health check: Sustainable development in the health and care system*. Cambridge: NHS England Publications.

Powells, G., H. Bulkeley, S. Bell, et al. 2014. Peak electricity demand and the flexibility of everyday life. *Geoforum* 55: 43–52.

Schatzki, T.R. 2010. *The timespace of human activity. On performance, society and history as indeterminate teleological events*. Lexington: Lexington Books.

———. 2016. *Multiplicity in social theory and practice ontology. Praxeological political analysis*. New York: Routledge.

Shove, E. 2009. Beyond the ABC: Climate change policy and theories of social change. *Environment & Planning A* 42: 1273–1285.

Shove, E., and G. Walker. 2010. Governing transitions in the sustainability of everyday life. *Research Policy* 39: 471–476.

———. 2014. What is energy for? Social practice and energy demand. *Theory, Culture & Society* 31: 41–58.

Shove, E., M. Pantzar, and M. Watson. 2012. *The dynamics of social practice: Everyday life and how it changes*. London: Sage.

Southerton, D. 2006. Analysing the temporal organization of daily life: Social constraints, practices and their allocation. *Sociology* 40: 435–454.

———. 2013. Habits, routines and temporalities of consumption: From individual behaviours to the reproduction of everyday practices. *Time & Society* 22: 335–355.

Southerton, D., A. McMeekin, and D. Evans. 2011. *International review of behaviour change initiatives: Climate change behaviours research programme*. Edinburgh: The Scottish Government.

Torriti, J. 2017. Understanding the timing of energy demand through time use data: Time of the day dependence of social practices. *Energy Research & Social Science* 25: 37–47.

Walker, G. 2014. The dynamics of energy demand: Change, rhythm and synchronicity. *Energy Research & Social Science* 1: 49–55.

Wallenborn, G. 2015. *The tragedy of energy efficiency. An interdisciplinary analysis of rebound effects*. Belgium: ULB – Universite Libre de Bruxelles.

Warde, A. 2005. Consumption and theories of practice. *Journal of Consumer Culture* 5: 131–153.

Zerubavel, E. 1979. *Patterns of time in hospital life: A sociological perspective*. Chicago: University of Chicago Press.

Stanley Blue is a Lecturer in Sociology at Lancaster University, UK. His work traces the reproduction of everyday practices that matter for sustainability and health. His current research examines the temporal organisation of working practices in large institutions.

Part 6

Researching Demand

15

Identifying Research Strategies and Methodological Priorities for the Study of Demanding Energy

Allison Hui, Rosie Day, and Gordon Walker

As the introduction established, there is considerable variation in accounts of energy demand. Starting from different places establishes different priorities and prompts different types of insights. This final chapter revisits the diverse chapters of the book with an explicit focus upon research strategies and how they relate to a shared interest in demanding energy. It recognises that explicitly discussing research tools and techniques can go some way to help stimulate what Mills (1970) might call the methodological imagination, and perhaps along the way make researching demanding energy less demanding for others. Explicitly attending to the research strategies used by the authors in this book is therefore part of developing a research agenda that places processes of demanding energy as its central concern.

The primary question underlying all of the studies discussed in this book is: what is energy for? This question builds upon the assumption

A. Hui (✉) • G. Walker
Lancaster University, Lancaster, UK

R. Day
University of Birmingham, Birmingham, UK

© The Author(s) 2018
A. Hui et al. (eds.), *Demanding Energy*, DOI 10.1007/978-3-319-61991-0_15

that energy is not used for its own sake, but in the course of performing a range of social practices that have distinctive spatio-temporal characteristics and dynamics of change. The social world, however, is vast, and given both the pervasiveness of energy and the on-going spread of energy infrastructures across many regions of the world, studying what energy is for could ostensibly lead to examining any part of social life. It is instructive therefore to reflect upon how the authors in this book have gone about their research, through composing research questions to address carefully defined concerns, delimiting suitable cases and selecting appropriate samples and methods.

The foregrounding of research strategies is deliberate. Whilst authors used a range of methods—interviews, ethnography, time use surveys, the analysis of secondary and visual data—these are not themselves particularly distinctive in relation to social scientific research in general, or energy research in particular. More innovative and consequential are the types of questions and aims that authors set out to investigate. These, in conjunction with the approaches to sampling and settings that follow from them, establish research designs probing aspects of demanding energy that have been understudied. This chapter thus begins by considering research questions and the priorities embedded in them. It then moves on to consider issues related to the selection of cases and samples, before concluding with some more general observations.

15.1 Methodological Priorities and Their Research Design Implications

While methodology and methods are often conflated in discussions of research, the former can be seen to involve not only techniques but also specific claims about the type of knowledge that is produced using particular techniques. Interviews or participant observation are therefore methods for gathering data, and thematic analysis through coding is one technique of working with that data. Thinking about methodology requires a broader view—a consideration of how the topic at hand is being approached and investigated, and what kind of knowledge is generated as a result. It is therefore about, for example, how research questions are phrased, which units of study are chosen and what research designs are able to investigate.

In the introduction to this book we argued that starting from a concern for what energy is for generates different understandings of how energy demand is constituted, patterned and changing. That is, it generates different knowledge about demanding energy. We are now in a position to reflect upon that argument in terms of methodology—if studying demanding energy generates new types of knowledge then this is in part due to particular methodological priorities. Another way of conveying this is that pursuing the question 'what is energy for?' involves a methodological approach that prioritises specific things during the framing of research questions, the defining of units of study and the formulation of research designs. Together, these priorities facilitate the production of a different type of knowledge about demanding energy.

Three methodological priorities that are embedded in this collection are outlined below. The specific formulation is ours—authors may not have engaged with these priorities explicitly within their own projects, but we argue that their work attends to them nonetheless. Nor is this list intended to be exhaustive—other shared priorities may also be found. The three methodological priorities outlined here, however, can be seen as points of intersection and agreement between authors, and provide a means of considering how the insights from one case might be considered in relation to other cases. They can also be read as prompts for future research—priorities that other researchers might embed into their own research questions and designs, and that policymakers might consider when looking for evidence to support particular interventions.

Priority one: it is crucial to pose questions that focus upon social dynamics rather than upon energy itself. Examples within this collection include: How is eating changing over time? What practices make up temporary events like music festivals? How is work within different departments in a hospital interconnected? How does leisure travel fit into the lives of diverse older people? Such questions are of significant relevance for discussions of past, present or projected energy demand, but they seek to first understand practices on their own terms. This is important because it acknowledges that everyday practices are organised around a wide variety of aims and goals (see discussion of timespace and human activity in

Schatzki 2010), aims that are often completely unrelated to energy for those engaged in them. One does not stay in a hotel in order to support energy demand management efforts—one stays at a hotel because of being away from home and needing somewhere to sleep overnight. Nevertheless, staying in a hotel has an impact on energy demand, its spatiality and temporality, and in aggregate, this is a significant effect. Collecting more detailed data on what is happening in the social world is thus crucial, and can reveal complexities and variations in processes of demanding energy that have been overlooked in the generation of existing projections, policies and paths for future change. Rather than assuming that cooking and car driving are performed similarly across a population, have universal meanings, or might be changed using similar strategies (e.g. campaigns asking people to make different choices), approaching them as topics of study brings into empirical question the extent to which they vary and the complex interactions that affect their transformation. Directing attention away from energy thus opens up opportunities for rich understandings of how social life is organised, and as a result paves the way for more sophisticated understandings of how demanding energy is a part of it.

Priority two: it is important to reflect upon how particular units of study facilitate the examination of different types of interconnections—for example within and between practices, technologies or organisations. Whilst the study of what energy is for can draw upon theories of practice, it is not synonymous with the study of practices (*praktik*) as understood within this literature (e.g. Reckwitz 2002). As we noted in the introduction, some of the authors in this collection do not draw in detail upon theories of practice, but rather build upon a broader social scientific literature interested in what people do. Whilst the concepts developed within theories of practice are particularly useful for developing understandings of what energy is for, they are not the only relevant ones. In addition to this general point, however, is a more specific one. To date, many studies applying theories of practice to the study of sustainability, climate change and related energy issues have been concerned with defining specific practices made up of constituent elements—for example using Shove et al.'s three-element model (2012). This is not, however, a necessary part of research designs. Starting out by defining practices of interest is only one of many research strategies available to those working more closely with theories of practice.

As our authors demonstrate, we can learn just as much about what energy is for by looking at how technologies are connected with each other and integrated into people's activities. Whilst at first glance it may seem this replicates the interest in technologies found in other energy literature, there is a key difference from much of this work. Rather than taking particular technologies or infrastructures as a primary focus, as found in user adoption studies or discussions of energy efficiency, authors interrogate how their interconnection with other technologies affects the process of demanding energy. For example, Wiig is interested not simply in mobile phones or tablets, but in how these are connected to infrastructures that facilitate their charging and mobile connectivity. He also investigates how infrastructures are themselves interwoven—with a provider of rail transport becoming increasingly involved in the provision of digital connectivity. A further example is Sahakian, whose study of household practices positions appliances as points of intersection between affluent householders, hired domestic helpers and architects or designers who each affect, in different ways, 'normal' practices. Curtis et al.'s discussion of demand management in hotels similarly considers not only the technologies and automated processes that make demand management schemes possible, but also how these materials connect to a series of activities—of which some are automated whilst others require the input of a range of employees in multiple sites. The use of one technology is therefore always connected, theoretically and empirically, to the use of other technologies as part of a wider system of practices that demands energy. As these chapters demonstrate, rather than asking questions about how technologies can be made more energy efficient or how more sustainable technologies can be more widely adopted, studying demanding energy involves asking different questions: How are norms around the use of technologies developed and reproduced? How do groups of people develop, maintain and use technologies in conjunction with particular infrastructures? How do multiple infrastructures interrelate materially and socially? The implication of these questions is that technologies have no unique importance within processes of demanding energy—rather, they are part of both infrastructural and normative systems that need to be empirically studied. Research questions, empirical data and analyses must therefore seek to query and retain at least some these connections.

What energy is for can also be investigated by focusing empirical work on organisational units and then investigating some of the many practices that occur within them. This approach provides not only an opportunity to consider how multiple practices are interconnected, but also to push against the segmentation of practices into discrete domains—as with the isolation of transport research from energy research. Many things take place in organisations, and starting from these units can prompt different kinds of insights about what energy is for and how demanding energy takes place. Blue's research focuses upon the work done within hospital settings, taking as a starting point the idea that interconnections between different departments and divisions are extremely consequential for understanding how patient care is delivered, and how energy is demanded as a consequence. Similarly, Jones et al. demonstrate in their study of construction and engineering consulting firms that the procuring and arranging of work within a firm has significant implications for the arrangement of 'necessary' travel. Taking a slightly different focus, Allen shows how 'events' can be seen as composed of a variety of practices. By thinking through live music events in rural and urban contexts, he raises important questions about how processes of demanding energy are situated in relation to diverse infrastructures. We might even think about the household as a kind of organisational unit, within which the diverse practices addressed by Greene and by Douzou and Beillan take place. In these studies, focusing on organisational units is thus not about providing analyses of those units per se. Instead, it is about thinking through other types of questions: How do dependencies and connections between departments affect the flow of work (and of energy)? How does the changing of one set of organisational practices affect other practices (and how they demand energy)? What are the implications for energy demand when organisations and events take on new practices that were not previously their concern (as with 'glamping' at music festivals)? How do the practices of household units bring into question the boundaries of home spaces? These questions recognise that the process of demanding energy is shaped by a range of organisations that structure work, generate expectations for co-presence and travel, coordinate infrastructures and inhabit varied spaces, among other things. Research questions, empirical data and analyses would therefore helpfully address and acknowledge these links.

Whilst different research questions are generated by approaching the study of what energy is for in each of these ways—from the definition of key practices of interest, from the systemic connections of particular types of technologies and from the designation of particular organisational units of interest—it is also possible to see how these approaches might also provide complementary insights. For example, Blue's discussion of CT scanner technology might have also been taken as the starting point for a more technology-focused investigation akin to that found in Wiig's chapter. Conversely, it would be possible to undertake a more institutionally focused study of Amtrak, the train company discussed in Wiig's chapter, and how the provision of digital connectivity, in addition to mobility, became part of its mandate. In this way the approaches used by different authors in the book offer possibilities for further investigating the range of cases presented here, as well as many others.

Priority three: whether research designs privilege practices, technologies or organisational units, it is methodologically important that they incorporate spatial and temporal dynamics. This is because what energy is for changes. Demanding energy is constituted by processes situated in specific locations marked by particular material and temporal relationships. The proximity of co-workers, the temporariness of infrastructures, the fluctuations of seasonal holidays, the rhythms of days and life cycles all affect how practices are performed and how energy is demanded. It is therefore important to build in opportunities for these dynamics to be recorded and analysed through research designs that probe variations and encourage comparisons across space, over time and in relation to practice-specific transformations. Whilst all of the chapters speak to these dynamics in some manner, those by Greene, Durand-Daubin and Anderson, and Jones et al. are particularly interesting for how they engage with trajectories of change in relation to different units—people, practices and institutions, respectively. Their work highlights how different trajectories of change are simultaneously shaping processes of demanding energy—meaning as a result that attempts to shape or steer future change will also encounter these different dynamics. Whilst this is a complex process, collecting more data that facilitates a consideration of units in relation to multiple, evolving spatial and temporal dynamics will provide a valuable basis of evidence. Research designs can therefore usefully consider questions such as: How do particular spatial relations matter to what energy is for? What temporal relations and variations can be observed? What exactly is changing over time and how might it be documented?

These three methodological priorities are embedded, in different ways, into the more specific questions that authors address in their empirical research. Whilst broadly cast, we argue that together they present a distinct approach that distinguishes the study of demanding energy from existing studies of energy demand. As such, they may be usefully taken up and developed by others interested in exploring new approaches to the study of energy demand.

Whilst these priorities end up being embedded into core research questions and the specifics of research designs, there are other techniques and tools that are also important within research strategies. The next section addresses one set—related to cases and sampling—in more detail, as this aspect of research design is very important for creating links between energy and social practices.

15.2   Approaching Cases and Sampling

Whilst asking what energy is for can prompt research questions that direct attention away from energy itself, it is of course important that it is not forgotten. We suggest that one of the aspects of research design where existing knowledge about energy demand is most central is in the specification and sampling of cases. As noted earlier, nearly any social process or practice could be studied and then connected, in some way, to energy. Attending to some cases and not others can therefore be justified on the basis of their importance for demanding energy.

Working through case studies in the first place could, in some energy research communities, be seen as a weakness. Positivistic approaches to research remain prevalent in non-social scientific disciplines, and therefore the acceptance of case study research that exists within the social sciences can be brought into question within interdisciplinary discussions of energy demand. However, in his powerful defence of case studies, Flyvbjerg notes that their utility is not limited to generating hypothesis or providing pilot insights, despite this widespread impression (2001: 77). Rather, he argues, careful sampling can make it possible to theorise on the basis of single case studies. What he calls information-oriented selection of samples and cases, based on expectations about the type of data they will provide, can be used to carefully address gaps in knowledge and provide important insights. In particular, he names three types of information-oriented selection that are particularly valuable when studying what energy is for:

1. Extreme/deviant cases: To obtain information on unusual cases, which can be especially problematic or especially good in a more closely defined sense.
2. Maximum variation cases: To obtain information about the significance of various circumstances for case process and outcome[...]
3. Critical cases: To achieve information which permits logical deduction of the type, if this is (not) valid for this case, then it applies to all (no) cases. (Flyvbjerg 2001: 79)

Each of these types of cases provides a means of linking insights to other cases and contexts not on the basis of representativeness, but on the basis of other exemplary, relational or logical ties.

Whilst case-based studies of demanding energy may therefore depart from some of the selection and sampling procedures established in other niches of energy research, the insights and assumptions from existing studies can prove central to case selection. The first example of this can be seen in the extreme case of Sahakian's Geneva expats. The experiences of this group of participants are not likely to be familiar to or common amongst the wider population in Switzerland, or other developed nations. But in energy terms they are interesting precisely because they, and the houses they inhabit, are out of the ordinary. Given the increase over time in the number of appliances built into houses and used by households, looking at a community where multiple appliances facilitate household practices can be argued to provide important insights on a case that might be deemed particularly problematic in energy terms. Rather than just understanding this group, it offers the potential to consider how levels of 'normal' energy use are established and maintained, when these norms are excessive in comparison to many other groups and contexts. It is not necessary that such an extreme case of energy use should be treated as a possible future norm for everyone. Indeed, this kind of projection can be problematic because of how it ignores the diverse trajectories of change at work. Nonetheless, investigating an extreme case can make it easier to see some of the relations between everyday practices, infrastructures, technologies and social norms that are easily overlooked when studying more familiar contexts.

Whilst none of the chapters in this collection focus centrally upon maximum variation cases, we can see that this approach to selection is apparent in some of their samples. Day et al., for example, segment their participants into three groups of older people, in order to more carefully consider the variations within older populations and how this affects travel demand. As their data shows, there are considerable differences in the embodied experiences and care responsibilities of different groups of older people, and these are consequential for imagining future patterns of travel demand. Similarly, Burkinshaw's project involved the selection of participants in different professions (architects, academics, graphic designers, accountants, solicitors and university support staff) that were expected to have contrasting access to flexible working. Whilst in the end he found no evidence of differential flexible working policies amongst these groups, this negative result was itself useful in determining the meaningful types of variation for discussions of working and commuting. The interest of many authors in spatial and temporal comparisons might also be seen as an important version of this selection strategy. In some chapters—such as Durand-Daubin and Anderson's work—the extent of this variation is constrained to some extent by existing datasets. Allen's chapter, however, demonstrates that pointing to variation—through reference to a wide range of different types of events—can be conceptually useful even where it is not tied to new empirical data. Engaging with varied cases or samples can thus be positioned to contribute a range of important insights about demanding energy.

Finally, building upon established knowledge about energy demand is most crucial for the selection of critical cases, which can be understood as framed in relation to assumptions. If a prevailing assumption is not true in a case where it might be most expected, then it makes sense to conclude that it is likely not true anywhere. Understanding existing assumptions about energy demand, and what energy is for, is thus important for those seeking to study critical cases. Several chapters of this book connect with this approach. Mullen and Marsden, Burkinshaw and Day et al. all frame their contributions in relation to existing arguments or assumptions about travel and how practices of travel are changing. These projections and suggestions about future change have been formulated in other energy research and policy contexts, but without detailed attention to whether they appropriately represent social dynamics. Therefore our authors, having chosen cases that put these assumptions to the test, are able to provide new insights that effectively dismantle them. Along the way, they highlight additional dynamics that need to be more carefully considered when developing future policies: the effect insecure working conditions have upon travel patterns; the interlinking of commuting with other types of travel and family practices; the importance of ageing bodies when considering leisure travel. In a different way, Douzou and Beillan, and Psarikidou, also position their arguments in relation to existing assumptions related to energy demand. Yet the assumptions they take on are not about how demand will or could change, but rather about its spatial and temporal characteristics. As a result, they are able to highlight how spatio-temporal meanings are quite complex. This complexity sits in contrast to (and indeed challenges) research using rational choice models that emphasise only the 'correct' attitudes and understandings that prompt choices aligned with sustainability or energy management objectives. Selecting critical cases thus provides a means of questioning and challenging existing assumptions about energy demand, whilst also generating data focused more upon demanding energy.

As these examples illustrate, careful consideration of case selection can ensure that even small studies or single cases can provide insights of relevance to broader considerations of demanding energy. In line with the second research priority above, cases may be of practices, of specific sets of people, perhaps in specific places or circumstances, of sets of technologies, or of organisations or settings. In conjunction with carefully formulated research questions, cases can be used to test or enter into dialogue with existing energy demand research, providing rich data about the social practices, spatio-temporal relations and transformations that affect what energy is for.

15.3   Conclusion

In researching processes of demanding energy, the authors in this collection have posed questions that foreground different concerns and enact different boundaries than other studies. For those working within particular empirical contexts or addressing energy demand in particular industries and policy realms, this will be apparent in terms of what is and isn't discussed and the specific insights provided by each chapter. This final chapter, however, has also demonstrated that these chapters have been marked by a distinct research approach. Starting from the question of what energy is for leads to different types of research questions that interrogate varied, interconnected units—practices, technologies, institutions and social groups—and how they are positioned in relation to space, time and change. Most centrally, this research strategy challenges the idea that energy research must place energy at the centre of all of its research questions; see in this respect a recent debate about whether or not social science research on energy should be expected to engage explicitly with energy units and measures in order to achieve policy relevance (Cooper 2017; Castree and Watt 2017). Authors have instead demonstrated that asking relevant questions about social life, in relation to carefully selected cases, can provide valuable understandings of how demanding energy is constituted, patterned and changing.

Opening up a discussion about research strategies has highlighted that researching what energy is for depends less upon particular methods than upon ensuring that the generation of research questions and selection of cases and samples is theoretically engaged and closely informed by an interest in social practices on their own terms. Starting from this point creates opportunities to develop more sophisticated representations of how demanding energy works—ones that for example can narratively connect the activities of those using mobile devices with the activities of those developing digital infrastructures to support them. On one hand this collection has exemplified how specific projects, with carefully positioned research designs, can contribute to new insights about what energy is for. Yet we also hope that it has demonstrated the importance of considering how multiple projects (as well as multiple practices, technologies, institutions and groups of people) can be looked at together. Better understandings of social life are developed incrementally. Developing innovative approaches to how we might bring together understandings and knowledge is therefore an important part of the creativity of research. By addressing a broad range of cases, and drawing out some of the points of intersection between them, we hope to have provided fodder for imagination and encouraged further efforts to produce innovative future accounts of what energy is for and how this is changing.

Bibliography

Castree, N., and G. Watt. 2017. What kind of socio-technical research for what sort of influence on energy policy? *Energy Research and Social Science* 26: 87–90.

Cooper, A.C.G. 2017. Building physics into the social: Enhancing the policy impact of energy studies and energy social science research. *Energy Research and Social Science* 26: 80–86.

Flyvbjerg, B. 2001. *Making social science matter: Why social inquiry fails and how it can succeed again.* Cambridge: Cambridge University Press.

Mills, C.W. 1970. *The sociological imagination.* Harmondsworth: Penguin.

Reckwitz, A. 2002. Toward a theory of social practices: A development in culturalist theorizing. *European Journal of Social Theory* 5: 243–263.

Schatzki, T.R. 2010. *The timespace of human activity: On performance, society, and history as indeterminate teleological events.* Lanham: Lexington Books.

Shove, E., M. Pantzar, and M. Watson. 2012. *The dynamics of social practice: Everyday life and how it changes.* London: Sage.

Allison Hui is an Academic Fellow in Sociology at the DEMAND (Dynamics of Energy, Mobility and Demand) Research Centre at Lancaster University, UK. Her research examines transformations in everyday life in the context of changing global mobilities, focusing particularly on theorising social practices, consumption and travel. Recent co-edited publications include *The Nexus of Practices* (Routledge, 2017) and *Traces of a Mobile Field* (Routledge, 2017).

Rosie Day is a Senior Lecturer in Human Geography at the University of Birmingham, UK. Her research interests centre around social inequalities in access to and experience of environmental and energy resources, with a related interest in environments of ageing and older age. She works collaboratively on a number of multidisciplinary projects in diverse international contexts.

Gordon Walker is Co-Director of the DEMAND (Dynamics of Energy, Mobility and Demand) Research Centre at Lancaster University and Professor in the Lancaster Environment Centre. He has wide-ranging expertise on the social and spatial dimensions of sustainable energy issues, sustainable social practices and cross cutting issues and theories of climate, energy and environmental justice.

Index[1]

---

[1]Note: Page numbers followed by 'n' refer to notes

© The Author(s) 2018
A. Hui et al. (eds.), *Demanding Energy*, DOI 10.1007/978-3-319-61991-0